WASTE PROBLEMS AND MANAGEMENT IN DEVELOPING COUNTRIES

WASTE PROBLEMS AND MANAGEMENT IN DEVELOPING COUNTRIES

Prof S K Singh

Delve PUBLISHING

www.delvepublishing.com

Waste Problems and Management in Developing Countries

Prof S K Singh

Delve Publishing
224 Shoreacres Road
Burlington, ON L7L 2H2
Canada
www.delvepublishing.com
Email: orders@arclereducation.com

ISBN: 978-1-77469-921-8 (Hardcover)

ABOUT THE AUTHOR

Prof. S. K. Singh is the Vice Chancellor of Rajasthan Technical University Kota and a former Professor & Dean at Delhi Technological University. He has a distinguished academic background, with a Ph.D. from BITS Pilani, M. Tech. from IIT BHU Varanasi, and B.E. from Gorakhpur University. He has over 35 years of experience in teaching, research, administration, and consultancy. Prof. Singh has published over 266 research papers, authored 10 books, and received 3 patents in the field of Environmental Engineering. He has guided 12 Ph.D. students and around 80 M.Tech theses. Prof. Singh has held various positions of responsibility, including Director of WAPCOS Limited and Head of Environmental Engineering Department at Delhi Technological University. He has received numerous awards and honors for his contributions to environmental conservation and protection. Prof. Singh has served on expert panels and committees for tasks such as evaluating projects, determining polluting industries, and selecting faculty members. He is also a member of various boards and governing bodies in prestigious institutions as well as a fellow of several professional bodies.

TABLE OF CONTENTS

LIST OF FIGURES

PREFACE

Waste is a severe issue since it has an impact on every form of biological existence. Water, air, and soil contamination are environmental problems brought on by wastes that endanger human health. The dangerous physiochemical consequences of wastes on biota, particularly in fresh water and terrestrial living organisms, have been elaborated on in recent works. In both the developed and developing worlds, the rate at which materials, energy, and other resources are being consumed to maintain the manufacturing of electrical and electronic items is obviously unsustainable. The use of mobile phones, personal computers, and flat-screen TVs has increased dramatically in recent years. New legislation has been introduced globally to encourage considerably higher rates of recycling and recovery of the materials used in the production of electrical and electronic products, whether they be metals, plastics, or organic chemicals.

The design of electrical and electronic devices is receiving a lot of attention in order to assist their treatment at the end of their life due to the technological complexity of much of the so-called e-waste and the challenges this creates for the recovery and re-use of precious resources. With the most recent scientific and technological advancements, enormous amounts of biosolid waste are being created all over the world, posing major risks to people, the environment, and agricultural fields. Along with many other benefits, careful management of biosolids can help reduce the load on landfills because they are rich sources of energy and nutrients. The want for protection against the health risks and environmental shocks of improper handling of hazardous waste increases along with the demand for a clean, safe environment. Entrepreneurs in business and industry show every day how people are becoming more aware of the need for improved waste management and the need to reduce or eliminate the sources of hazardous waste. Heavy metal pollution is created by industries like tanning, pulp and paper, sugar, fertilizer, and textiles. This pollution has an enhanced negative impact on the environment, especially on water streams, and it can lead to significant ailments. Utilizing radioactive materials for nuclear power generation in reactors and the development of nuclear weapons results in the production of radioactive waste. The environment and the various forms of life that currently exist on this planet are seriously threatened by the use of nuclear fission reactions and other types of nuclear applications in the realms of medicine and research. Since the last few decades, the use of radioactive elements has increased yearly, causing an enormous buildup of radioactive waste in the environment.

Introduction

Municipal solid waste management (MSWM) is one area of infrastructure under tremendous demand due to rapid urbanization in cities. The municipal solid waste is increasing, and the infrastructures that are currently in place cannot handle it. The main cause of inadequate project planning and regulation implementation in developing countries is a lack of understanding of the different technologies available, the best management practices, and the foundations of MSWM. It is required to establish suitable waste collection, treatment, recycling, and disposal procedures as well as an understanding of various treatment technologies in order to achieve a sustainable MSWM system in a developing country. The development of an effective system necessitates the inclusion of all technical, political, environmental, socioeconomic, legal, and cultural aspects of planning and developing an integrated solid waste management (ISWM) system. Future environmental technologists and managers have been educated through a basic skills curriculum at the undergraduate level, followed by a graduate program in engineering or a science field.

Large dumps have proliferated all over the surrounding areas of major cities as a result of the rising waste generation. This solid waste represents a significant loss of both material and energy resources. They can also present a variety of direct and indirect threats to the environment and human health if they are not properly handled. These risks include air pollution brought on by greenhouse gas emissions, infectious diseases, or contamination of the soil, groundwater, and surface waters. Depending on their socioeconomic development, the money earned from recovering garbage could provide these countries with considerable economic opportunity. The important factors contributing to the rise in waste are the emigration from rural areas, the growth of the waste industry, and changes in household consumption patterns. Despite the fact that most cities have mechanisms in place for collecting waste, their removal reveals a management gap because most landfills are unregulated and have a negative impact on the environment.

This book explains the importance of waste and its management. It can serve as a hand book for students and researchers.

CHAPTER 1

Introduction to Waste Management

CONTENTS

1.1 INTRODUCTION

With the growth of cities and the population, waste generation, collection, and disposal became a p roblem (Aryampa, Maheshwari, Sabiiti, Bateganya, & Bukenya, 2019). A large percentage of the budget is spent on waste management (WM). The cities are faced with important issues due to the lack of resources.

Figure 1.1. Waste management. (Source: Van Thanh N. Optimal Waste-to-Energy Strategy Assisted by Fuzzy MCDM Model for Sustainable Solid Waste Management. *Sustainability*. 2022; 14(11):6565. https://doi.org/10.3390/su141k.jpg, Creative Commons License).

Despite the efforts of the government, non-governmental organizations, and commercial companies in the WM sector, the situation has not entirely improvet WM) (Figure 1.1) refers to any procedures and techniques used to reduce, recycle, reclaim, or get rid of waste that is often produced by human activities (Allesch & Brunner, 2015). It refers to all activities that are done to decrease their negative effects on human health, including prevention, pre-collection, collection, transportation, sorting, reuse, disposal, and/or treatment. The primary challenge for the WM sector is adjusting to this rising urbanization rate and striking a balance between service quality and cost-effectiveness. Another challenge is to examine the difficulties, restrictions, and constraints of the WM experiences that have already been had in developing nations.

1.2 DATA ANALYSIS

The first step towards enhancing this industry may be to understand the causes of the mistakes in this data. Before beginning to plan a garbage collecting system, for instance, it is imperative to obtain accurate data on waste density. As a result, it's crucial to concentrate on choosing the greatest possible WM option using the findings from the trustworthy data that will be gathered. The key to designing a solid waste collection system and deciding on the best alternatives for treatment and recovery will be an accurate understanding of these indications and the waste composition. In order to choose the best integration strategies for the informal sector, it will be crucial to use the data on the sector that has been collected. This information will also be helpful for managing, and recovering municipal solid waste. We can use all collected data to determine the best method for trash collection, treatment, and recovery. The availability of collected data and end-user knowledge management techniques can offer effective WM in both cities (Ferraris, Mazzoleni, Devalle, & Couturier, 2018)thereby improving performance. However, the management of the knowledge generated from the BDA as well as its integration and combination with firm knowledge have scarcely been investigated, despite an emergent need of a structured and integrated approach. The paper aims to discuss these issues. Design/methodology/approach Through an empirical analysis based on structural equation modelling with data collected from 88 Italian SMEs, the authors tested if BDA capabilities have a positive impact on firm performances, as well as the mediator effect of knowledge management (KM. As a result, we will achieve both the mitigation of climate change and the minimization of air, water, and land pollution. The increase in recycling rates will also help to generate a large number of new jobs. By properly integrating the formal sector, the workers in the unorganized sector will become more organized and earn more money, which will help to reduce poverty.

Figure 1.2. Municipal solid waste (MSW) management. (Source: https://upload.wikimedia.org/wikipedia/commons/f/f8/Rifiuti_solidi_urbani_%28Municipal_solid_waste%29_-_Stazione_di_Aversa_-_Province_of_Caserta%2C_Italy_-_11_Feb._2008.jpg, Creative Commons License).

All municipal solid waste (MSW) management (Figure 1.2) activities require the use of appropriate infrastructure, services, and upgrades m, & Kothari, 2011)which generate thousands tons of municipal solid waste daily, which is one of the important contributors for environmental degradation at national level. Improper management of municipal solid waste (MSW. This is becoming more costly and difficult as a result of the unchecked and continual growth of urban areas.

Urban communities struggle to provide the amount of public service necessary due to their precarious financial status. Additionally, the planning, designing, and operation of the management system are all based on the type and quantity of MSW produced.

The physical and chemical characteristics of MSW have been demonstrated to vary with population density; these differences reflect the effects of urbanization and development. Three techniques are widely used to describe waste: direct sampling, market product analysis, and waste product analysis. Direct sampling is useful in the absence of site-specific data, but waste product analysis and market product analysis are appropriate for large geographic areas.

1.3 MSW

For instance, some nations' extensive use of vegetables for a variety of purposes results in a high organic content in MSW (Shiralipour, McConnell, & Smith, 1992). It is obvious that some cities generate a higher proportion of organic matter than developed nations.

The proportion of organic waste grows proportionately as a country's income declines. Seasons, way of life, dietary choices, and level of commercial activity are some of the factors that affect how much MSW is created. It is possible to better plan collection and disposal systems by using data on quantity variation and generation. Rural families produce more organic garbage than urban households do because the proportionate percentage of organic waste in MSW frequently rises with declining socioeconomic standing. The region with the lowest income level had a higher proportion of organic garbage than other locations. The area with the lowest income level had the highest amount of organic waste, whereas the area with the highest income level had the lowest rate. Food and kitchen waste accounted for the majority of the total waste in HWs across all socioeconomic groups. However, it was discovered that income had no effect on the overall amount of municipal garbage/waste generated during the research that was done to know the relationship between waste generation and socioeconomic characteristics, such as income and population density. Furthermore, the season has a big impact on the composition of municipal waste. In comparison to waste made of cardboard, fiber, bottles, wood, metal, and glass in the winter, organic waste is more frequent in the summer.

Kitchen garbage (Figure 1.3) is the component of municipal solid waste that differs most between the summer and winter seasons. This change is assumed to be the result of water from solid waste evaporating due to summer's high temperatures and increasing fruit consumption. Given the variations in organic waste according to income level and seasonal conditions, additional modifications relating to income level were discovered in the winter compared to the summer. In this context, MSW categorization research could be based on two methods: on seasonal fluctuations and on various income categories (Kamran, Chaudhry, & Batool, 2015). One of the biggest challenges that public agencies face is waste collection. These issues frequently lead to the piling of waste and the development of unregulated open landfills. MSW collection is an important and expensive public service that normally makes up between 75 and 80 percent of the MSW management budget.

Figure 1.3. Kitchen garbage. (Source: https://upload.wikimedia.org/wikipedia/commons/b/b7/Corroded_Garbage_Disposal_From_Above.JPG, Creative Commons License).

1.4 WASTE COLLECTION

Pre-collection and collecting of waste are both included in waste collection. Some research studies discuss solid waste collection in low- and middle-income countries in detail, despite the importance of collection and pre-collection techniques. Observe that poor collection continues to exist in many low-income cities. It is well-known that the degree of the collection rate and the income of the nation are correlated. Waste collection services may be provided by the formal sector, through public or private sector operators, or by the unofficial sector, for instance, through community-based organizations (CBOs), non-governmental organizations (NGOs), or micro- and small businesses (MSEs). Collection services offer a primary collection to small-scale local neighborhoods or a secondary collection service to the entire city. Waste pre-collection refers to all procedures for removing it from the production site. All procedures that come before the collection process are implied by the term "pre-collection." The collection is the sum of all the actions that include removing garbage/waste from places of regrouping and sending it to a location for sorting, regrouping, recovery, treatment, or storage. It entails gathering garbage for transportation. There are two types of collection: (1) door-to-door, in which the collection service makes sure that garbage is disposed of regularly; and (2) voluntary drop-off, in which the producer delivers waste to a collection location so that it can be transported to a disposal or treatment facility. A relatively sophisticated system, waste collection consists of moving, interconnected containers and vehicles(Gentil et al., 2010).

The number of vehicles utilized for waste collection, the efficiency of bin collection systems, poor route planning, lack of knowledge about collection schedules, inadequate infrastructure, terrible roads, and insufficient infrastructure all have an impact on collection methods. Sometimes it is necessary to reassemble the garbage through a transfer station before it is transported to the center of treatment, usually for financial reasons. In transport, regrouping, and waste collecting hubs, this temporary storage is possible. It is crucial to monitor the waste conditioning during this storage phase and take various precautions to reduce the environmental impact while awaiting transport and treatment. Waste disposal in poor nations is frequently restricted to initial collection by organizations like NGOs. The lack of appropriate and functional trucks and equipment makes it difficult for municipal technical services to perform secondary collection, which is frequently their obligation. The prevalence of manual handling activities that expose workers to sanitary health risks is another factor that makes the waste collection process more crucial in developing nations. The collection system includes various collecting techniques that describe how the container interacts with the worker and the vehicle. Manual gathering is a manual collection process, the worker carries, lifts, and loads waste containers or bags onto the vehicles. It is a combination of mechanical and human processes, with the truck acting as the only mechanical component for lifting and emptying containers close to the vehicle. Although the semi-automated collection process is entirely automated, a person is still required to operate the coupling and provide human assistance during the coupling and uncoupling of the vehicle-container. In automated collection, a single operator within the vehicle cabin manages the communication between the container and the truck (Figure 1.4) during completely automated collection, which restricts workers' direct involvement.

In terms of administration, there are two ways to handle waste collection. Direct collection is the method where the local authority decides to internalize the service by bearing the cost of equipment and its replacement (bins, trucks, etc.), as well as the operation of the service (personnel, territory), and delegated collection, where the design, implementation, and operation of the service are carried out by external companies, as permitted by the public procurement code. In many African cities, the percentage of waste collection is under 50%, which is extremely low (Kassim & Ali, 2006)in both developed and developing countries. People always generate solid waste through their daily activities.

Figure 1.4. Waste collection truck. (Source: https://upload.wikimedia.org/wikipedia/commons/7/71/Waste_collection_truck.jpg, Creative Commons License).

This solid waste needs to be properly managed in a way that minimises risk to the environment and human health, which means storage, collection and proper disposal. At the same time solid waste creates livelihoods for the urban poor in terms of employment and business. This paper looks into one aspect of SWM, namely collection service. The importance of SWM is now recognised at international, national and community level. The Agenda 21 declaration of the United Nations [UN. (1993. In order to create a criticality index for garbage/waste collection, it is necessary to first identify the major variables that affect waste management, such as waste generation, inadequate collection services, and poor urban infrastructure. The criticality index assists in emphasizing these important components and highlights the difficulties in obtaining accurate data on waste management in African cities. However, the efficiency of solid waste collection is constrained by corruption and a lack of openness in both the public and commercial sectors. By adjusting private sector engagement to socioeconomic realities, stepping up efforts to promote competition, and putting policies in place to guarantee that low-income households have access to affordable services, the problems with service provision in low-income nations may be handled. The public and private sectors must work together in a formalized public-private partnership to achieve this.

Source-separated collection of MSW generated by households is one of the crucial steps for integrated solid waste management (Tai, Zhang, Che, & Feng, 2011). Unfortunately, most cities have yet to successfully implement source-separated collection at the household level, which is the first stage of the waste management cycle. The development of a solid waste management system that is effective and environmentally responsible is a major problem for economies that are expanding. The collection and transportation operations account for between 80 and 95 percent of the total budget, which largely determines the economics of the entire MSW system. Poor design of any of these processes may result in harm to the environment and increased operating expenses. Ineffective solid waste transportation will have an influence on management services since it will increase operational expenses and diminish earnings. Municipal agencies employ their own vehicles to transport MSW, but in some locations, they also hire outside contractors. Transport is by far the biggest and most expensive part of solid waste management because of how labor-intensive the operation is and how frequently trucks are used(Tin, Wise, Su, Reutergardh, & Lee, 1995). Unfortunately, transportation of solid waste is a problem in many developing nations. It was determined that local authorities' finances are severely strained by the high cost of waste collection and transportation as well as the challenges associated with handling solid waste. It is more challenging to provide adequate solid waste collection and disposal services because of a lack of funding, resources, customer opposition to paying, and the wrong use of economic tools. Legislators sometimes give other municipal duties priority over solid waste management, which leaves local governments with a shortage of qualified and skilled staff. Furthermore, the low social status situations in which workers are involved demotivate them from giving their best effort. Competence must take precedence over preference and this dedication must significantly improve in order to ensure the safe collection and disposal of solid waste.

1.5 TRANSPORTATION

On collection and transportation operations, it has been discovered that poor collection schedules, improper bin collection systems, inadequate infrastructure, and a lack of trucks for garbage collection all play a role.

Important elements of the collection and transportation of solid waste are operations planning and vehicle routing. Numerous studies have indicated that in order to ensure efficient waste collection and transportation, planning should take into account systematic routing utilizing scientific approaches that integrate cost savings and environmental conservation. In poor nations, there is no coordinated or systematic procedure for allocating trucks or other vehicles for the collection of solid waste. This is typically based on practical experience and intuitive methods, which result in expensive and inefficient practices that have negative consequences on the environment, public health, business operations, and other aspects. Most of the cities lack a robust, adequate transportation infrastructure. Poor roads and a high vehicle population have a negative impact on the collection and transportation of solid waste (Hazra & Goel, 2009)India and suggests solutions to some of the major problems. More than 2920ton/d of solid waste are generated in the Kolkata Municipal Corporation (KMC. Both the nature of the roads and poor route planning during site physical development have an impact on effective collection. One technique to ensure better performance in the collection of solid waste is to route collection vehicles efficiently. The technical proficiency or competence of the operational crew also has an impact on WM. The WM system's technical aspects are related to a lack of knowledge. To operate at their best, operational staff should primarily be chosen based on their technical competence. The employees should be knowledgeable about the most recent difficulties and technological advancements.

These consequences are largely mitigated by reducing resource use through recycling and energy recovery through incineration. The results draw attention to the fact that underestimating the environmental impacts of a waste management system, particularly when it comes to the use of fossil fuels, and climate change, may result from ignoring the consequences of collection and transport. To lessen the environmental impact of waste management systems, accurate optimization of waste transit is required. Optimizing garbage transfer and transportation, which is the most expensive activity, requires the finest route planning for trucks (Nguyen-Trong et al, , 2017). Optimization of individual routes and scheduling for the entire collection period can result in significant cost savings compared to the norm. These options include alternate weekly collection, a cooperative collaboration between neighboring authorities, and placing vehicle depots closer to waste disposal facilities. Abundant research has been done to find the WM sector's shortcomings and issues. The majority of research has examined the waste characterization, creation, collection, storage, transportation, treatment, and reuse processes of WM systems. Even with all of the institutional actors' efforts to improve this industry and despite considerable developments, waste upstream and downstream management still has substantial problems.

Figure 1.5. Door-to-door waste collection. (Source: https://upload.wikimedia.org/wikipedia/commons/7/73/Door-to-door_Garbage_Collection_Vans_in_Indore_-_Under_Swachh_Bharat_Abhiyan.jpg, Creative Commons License).

In order to enhance the people's participation in the democratic process and control over activities that directly impact them, effective MSW management entails boosting access to information and its presentation to the public. The objective of successful MSW management includes working with homes and waste producers as well as those further upstream in the management cycle. It is an upstream/downstream relationship that is dynamic. Despite this, we can still see that there are only a few operationally feasible ways to eliminate MSW at the source. Feedback from this initial phase of the plastic bag ban law's implementation demonstrates that the consumer's continued importance as a major success component warrants special attention. Promotion of waste recovery, recycling, and sorting, identification of the unorganized sector and study of prospects for collaboration, inclusion of small businesses in waste management, and bolstering the supply of waste management in rural areas are other crucial MSW success aspects. Door-to-door collecting (Figure 1.5) is not used in the majority of Indian cities. The practice of classifying MSW at the household level does not exist. But in certain cities, private organizations collect recyclable rubbish, which is then mixed with other wastes in municipal bins and taken to disposal locations.

Due to the lack of sufficient facilities to treat and dispose of the greater volume of generated MSW, MSW management is currently in a critical stage. When waste is disposed of improperly, it has a negative impact on all elements of the ecosystem and human health. For all activities, the management of MSW necessitates sufficient infrastructure, maintenance, and upgrading. In metropolitan areas, the inability to provide the appropriate degree of public service is frequently linked to the controlling municipal corporations' precarious financial situation. Authorities are in charge of managing municipal waste on a general level and are dealing with a particularly challenging scenario when it comes to handling the waste that is produced. There are no controlled landfills in some countries, and the majority of this waste is still not managed and released in public areas. Some findings indicate that financial constraints, a lack of effective legislation, a lack of knowledge of the factors that influence the various steps of WM, a lack of authority commitment, and a lack of sufficient updated data on municipal solid WM systems are the main causes of this critical situation. For WM, an integrated strategy has been proposed. The major elements of this strategy are improving shareholder communication and integrating techniques and technology that are appropriate and readily available to fulfill the goals of an efficient WM system. The plan calls for encouraging nongovernmental organizations to use composting equipment to create biofertilizers and encourage people to start more recycling businesses.

1.6 WASTE PRODUCTION

Over 42 million m^3 of general wastes are produced annually in South Africa (Guo, Song, & Buhain, 2015). The huge volume of solid waste stream from mining, pulverized fuel, agricultural, and urban waste activities is the cause of this. South Africa produces more waste per person per day than industrialized nations like the UK and Singapore, 0.7, 0.73, and 0.87 kg/person/day, respectively, which is a dilemma for waste management in this nation. It is challenging to classify and describe waste collection systems without a common understanding of capabilities and features, which restricts not only the evaluation of operational elements but also the development of distinctive and sustainable garbage collecting systems. Outstanding achievements in solid waste management and resource total utilization have been made by modern industrialized nations. Between 1960 and 2004, solid waste management practices underwent considerable changes in many nations, including Germany, Sweden, Japan, and the USA (Karak, Bhagat, & Bhattacharyya, 2012). One significant development was the beginning of solid waste management with reduction, using less initially and reusing more, and recycling. Landfill disposal has been replaced as the major way to treat solid waste by composting and the burning of organic waste. Waste is both a resource and a risk, but when it is disposed of carelessly, it poses a threat to the environment, harms the environment, and exposes people to hazards, some of which can be extremely serious.

In addition, WM that is unreliable and out of control promotes the upansion of slums. A few landfills follow environmental standards that are recognised recognized in industrialized countries. In these dumps, gas emissions from decomposing waste pose a serious threat to the environment. Groundwater is a crucial source of drinkable water and waste disposal practices have the potential to contaminate it since leachate carrying various harmful compounds gradually seeps into the groundwater over time. The open-air combustion of waste deposits, which emits abundant volumes of smoke and foul aromas, is the main contributor to air pollution. The physicochemical and/or microbiological contamination of soils is caused by atmospheric emissions from incineration, the percolation of landfill leachate, and the spreading of components or sludge. The effective application of technological approaches, their adaptation to actual conditions, and the inclusion of unauthorized waste collectors and scavengers, however, could have a favorable effect on WM strategies. Scavengers and informal reuse collectors also provide demonstrable economic and environmental benefits to the community; as a result, their activities should be promoted and developed. The amount of waste is constantly rising in all global cities, whether they are in developed or developing nations. Limiting landfilling and incineration requires changing practices due to the exponential growth of wash is a result of urbanization and shifting consumer patterns, particularly in emergingdeveloping countries. Waste recovery is therefore becoming more and more crucial. This waste recovery, which is primarily a political and social decision, is not only beneficial but also desired. It entails, in particular, giving waste a market value and is accomplished through a variety of methods, resulting in raw material savings while also directly promoting environmental protection.

Figure 1.6. Compost. (Source: https://upload.wikimedia.org/wikipedia/commons/e/e2/Compost_made_from_faeces_and_kitchen_waste_%282947143382%29.jpg, Creative Commons License).

The influence on and preference for solid waste management are strongly correlated with a country's level of development. Compost (Figure 1.6) and other new goods, such as renewable energy, are produced by many affluent nations using various waste management strategies. These countries invest in waste recycling to help their agricultural industries. The selection of solid waste management will depend on decisions made by city authorities regarding the type, quantity, and quality of locally produced garbage as well as other variables. Investigations into classification and comtion are crucial to such waste management since they aid in selecting and preparing for the best system for the transportation, storage, disposal, and recovery of solid waste. Characterization is essential in intervening to pinpoint any potential environmental impacts on humankind and the environment. Even if there are still many obstacles to overcome, the recovery of waste has clearly improved in the Northern countries and, in particular, the countries of the European Union. European nations have created a strategy to maximize waste recovery using a variety of procedures.

Recycling has been successful in Sweden (Figure 1.7) recently, with 36% of garbage recycled, 14% composted, and, most importantly, 49% burned, the highest rate in the Union after Denmark (54%) and significantly higher than the European average (22%) (Malinauskaite et al., 2017). A novel approach to waste management based on the idea of integrated waste management is strongly advised because conventional SWM practices cause problems and in order to protect the environment and the health of the population from the perspective of economic, environmental, and sustainable development. Therefore, the idea of integrated waste management is strongly advised. Implementing strategies including source reduction, sorting, recycling, reuse, and regeneration is part of an integrated SWM strategy to waste management.

Figure 1.7. Recycling. (Source: https://upload.wikimedia.org/wikipedia/commons/5/5e/Paper_recycling_in_Sweden_2020.jpg, Creative Commons License).

1.7 RECYCLING

Recycling is the complete or partial replacement of a waste product with a new raw material in the production cycle from whence it originally came (Huntzinger & Eatmon, 2009)the third largest source of carbon emission in the United States. In addition to the generation of CO2 the cement manufacturing process produces millions of tons of the waste product cement kiln dust (CKD. Reuse is the new application of a waste product for a function related to its original usage. In a sense, it prolongs the product's life before it is wasted/destroyed. Fore, after cleaning, the bottle is restocked. This kind of recovery is frequently seen in emergingdeveloping nations in Africa. A waste product's qualities are restored through regeneration (valorization), a physical, chemical, or biological process that allows it to be employed as a replacement for fresh raw materials. Another waste management strategy is energy recovery, which involves using the calorific value of waste products to generate heat or electricity. Usually, methanization or incineration are used to accomplish this. Another way to manage waste/trash is through composting, particularly organic waste from forestry, animal husbandry, the food sector, and upkeep of green spaces. Municipal solid waste (MSW) typically has a calorific value between 465 and 6978 Kj/Kg and nitrogen and phosphorus concentrations between 0.5 and 0.8 (Karouach & El Bari, 2023)2023. Lack of finance, a lack of political will, bad scheduling and routing, inadequate road network and development planning, and subpar technological components are all barriers to improving WM performance. These issues are more noticeable in developing economies. Particularly in low-income nations, providing universal access to waste management services and preventing illegal trash/waste dumping are top priorities. Therefore, to find sustainable, context-specific solutions, municipal officials and local people must work together for effective waste management in these situations.

An improvement in the collecting and transportation networks is necessary for optimal solid WM. The disposal service is damaged by a lack of garbage collection and transport infrastructure. For instance, placing the bins distant from the sources of garbage generation will promote unethical behaviors like dumping trash in public places. Organizing the unorganized sector and encouraging suitable microbusinesses are efficient ways to enhance waste collection services. Expanding recycling and integrating it into the formal system presents a substantial win-win opportunity because recycling in the informal sector currently saves many cities millions of dollars and provides a means of support for a sizable section of the urban poor. This data is the consequence of incorrect and erroneous waste production data collection and reporting practices. Studies should focus on the availability of data, adequate system design, optimization of truck routes, and incorporation of informatics and mathematical programming to ensure resource conservation through the reduction of distance, time, traffic problems, and operational and maintenance costs. Solid waste management (SWM) systems in developing countries (DCs) are hampered by a number of issues, most notably a lack of technical expertise, a lack of public and administrative knowledge of environmental implications, and a lack of financial resources (Kaseva & Mbuligwe, 2005)which was carried out in Dar es Salaam city to assess post-privatisation of solid waste collection and disposal. Prior to the assessment, fieldwork studies indicated that current solid waste generation rate in the city is 0.4kg/cap/day and total waste generation is within the range of 2425tons/day. This study also indicated that out of the total waste generated, a total of 957tons/day is collected by the three city municipalities (231tons/day or equivalent to 10% of the total generation. It is also possible that international funding and collaborations, rather than improving the situation from a long-term perspective, may exacerbate it by implementing solutions that ignore local problems and instead end up importing overly complex solutions that fail quickly or that worsen social inequalities. When it exists, the local informal sector, for instance, collects and recycles valuable materials like metals, plastics, and glass. However, if it is not considered when a formal waste management plan is developed by local or international authorities, it will have a negative impact on the local economy.

Figure 1.8. Waste burning. (Source: https://upload.wikimedia.org/wikipedia/commons/9/94/Burning_Garbage_-_Padmapukur_Water_Treatment_Plant_Road_-_Howrah_20180528081344.jpg, Creative Commons License).

Garbage accumulates along streets, in open dumps, or at best, in subpar landfills as a result of an inefficient or poorly managed SWM system, typically in conjunction with community waste burning (Figure 1.8). Adopting sustainable solutions is the main objective in order to avoid the already severe situation growing worse and, whenever possible, to turn the need into an opportunity. This is due to the necessity of setting up a SWM system that is capable of successfully addressing and resolving pressing issues. The acceptable technologies should, to the extent possible, involve operators who have undergone particular training in addition to simple, readily available local options. Sanitation can improve living conditions by providing essential standards for waste disposal and cleanliness. Poor wastmanagement causes the public to come into direct contact with the waste, making them susceptible to viralus diseases, accidents, diarrhea, and respiratory issues. The already fragile situation will only get worse as a result of the contamination of the air from open burning, the surface and groundwater, and the soil as a result of leaching. A subsistence economy is one in which the use of the appropriate technologies, the creation of financial advantages and new business prospects, and the formalization and financing of informal sector activity make it possible to repurpose waste as a resource. The marity of waste pickers are unorganized workers who are not covered by social security or insurance programmes; they offer options for self-employment in extremely difficult working environments. Given the existence of the informal sector in this context, it is crucial.

Figure 1.9. Sustainable landfilling. (Source: https://upload.wikimedia.org/wikipedia/commons/2/2f/Landfill_area_background_-_Arri%C3%A8re-plan_de_l%27aire_d%27enfouissement_%283266498035%29.jpg, Creative Commons License).

1.8 LANDFILL

Sustainable landfilling (Figure 1.9) is a means to appropriately dispose of waste that has no practical or commercial value while breaking the cycle of materials that return minerals and carbon to the earth (Kale et al., 2007)56.3 million tons or 25% of the MSW in 2005 in Europe, and 3.3 million tons or 10% of the MSW in 2004 in Australia. Currently, in the USA the dominant method of packaging waste disposal is landfill, followed by recycling, incineration, and composting. Since landfill occupies valuable space and results in the generation of greenhouse gases and contaminants, recovery methods such as reuse, recycling and/or composting are encouraged as a way of reducing packaging waste disposal. Most of the common materials used in packaging (i.e., steel, aluminum, glass, paper, paperboard, plastics, and wood. Open dumps lack barriers to prevent leachate and biogas dispersion from harming the ecosystem. Waste pickers working at the disposal site are exposed to health risks due to open burning that generates toxic fumes and unregulated waste disposal. When constructing a sustainable landfill, long-term emission reduction and danger avoidance must be taken into account. In technologically and economically constrained environments, the following components should be present: low-cost solutions for development, operation, and maintenance; simple, easily implemented technologies; and maximum utilization of in-situ materials and natural resources. Landfilling has traditionally been the primary method of waste treatment, changing through time from an open pit to a sanitary landfill, progressively raising the level of technological complexity and environmental criteria. Any waste deposit on the ground functions like a typical three-phase reactor: water and air are introduced to the reactor as separate streams or as parts of the mass of waste that is being introduced. Leachate and biogas emissions are created in the landfill reactor by biological, physical, and chemical-physical processes that transform both organic and inorganic materials. Waste is the source of any potential residual emissions. Sustainable landfilling is becoming a fundamental idea that cannot be ignored while developing SWM systems. A sustainable landfill is one that should achieve an acceptable equilibrium with the ecosystem in 30 to 40 years (Spokas et al., 2006). It is possible to think of the acceptable equilibrium in terms of FSQ compliance and long-term emissions control. The term Final Storage Quality (FSQ) refers to a group of values for several parameters that must be attained over the course of one generation to indicate a suitable equilibrium between the environment and the landfill. According to FSQ, the factors that have the greatest long-term effects are the quality of deposited waste and emissions.

To meet sustainability goals, a landfill's design must be appropriate, and a combination of mechanical, chemical, biological, and thermal treatments must be used before waste is deposited, in situ during operations. These treatments must be aimed at promoting the stabilization of the waste, which lowers the long-term mobility of contaminants. Using a landfill, the term bioreactor was created expressly to describe a landfill where various in-situ processes are used to boost biological degradation or methane production, as well as boost ammonia and organics transformation. Leachate recirculation, the addition of water, and natural or forced aeration are a few examples of these approaches. Improved control of biochemical kinetics, moisture content, and redox conditions may decrease the pot treatment phase significantly and, as a result, hasten the realization of environmentally sustainable goals. By injecting air into the waste pile, the aerobic atmosphere inside the landfill may be maintained, boosting the growth of aerobic microorganisms and thereby the biodegradation processes, which are up to 10 times faster than anaerobic ones. Aeration of the waste material avoids the production of methane and, consequently, energy recovery; on the other hand, it lowers the leachate nitrogen and carbon values, saving money on leachate treatment. Forced aeration requires energy to maintain, and problems with efficient air circulation inside the mass might reduce the effectiveness of the aeration. More than a planned alternative for active landfill management, forced aeration is mostly employed for remediating old anaerobic landfills. By promoting the presence of oxygen in the waste mass without the use of sophisticated technological solutions, semi-aerobic landfill systems hasten the stabilization of waste and the removal of nitrogen from leachate. Semi-aerobic landfilling offers a beneficial compromise between the technological complexity and costs of aerobic systems and the long-term effects produced by anaerobic systems, even if it is characterized by a lower stabilization efficiency than aerobic systems. The natural ventilation technique is based on a simple physical idea: air flows into garbage/waste layers via a natural advection process caused by the temperature difference between landfill levels and outside air

The goal of hybrid bioreactors is to combine the benefits of both aerobic and anaerobic operating conditions to increase energy recovery and hasten waste stabilization (Ardakani & Badalians Gholikandi, 2020). The accumulation of ammonia in the landfill (aerobic-anaerobic systems) or the high expenses associated with constant air injection and leachate recirculation will depend on the combination. Relaying to various barriers is important in order to effectively avoid or mitigate uncontrolled emissions; any method created to lessen the mass transfer of toxins from the dump to the surrounding environment is referred to as a barrier. A crucial element in the construction of sustainable landfills is determining the best mix of the barriers. When it comes to the contemporary idea of sustainable landfilling, landfill can be seen as a cascade of many barriers. Emissions can be physically contained by barriers, or they can be positively impacted by controlling the transformation process inside the landfill's main body. Five barriers are represented by controllable valves based on how they perform. The top cover, lining systems, biogas, and leachate capture systems are examples of physical barriers that either ensure containment or can manage pollutants by allowing them to drain out fast in the direction of final treatments. However, the reactor barriers encourage biological processes and the transformation of carbon, and depending on how landfills are managed, the generation of leachate or biogas may be increased. Increased rates of waste output in expanding, already overcrowded cities are contributing to an increase in the number and seizure of open landfills. By 2050, the world population is projected to increase from over 50% to 70%, with the majority of that expansion taking place in the most impoverished regions of Asia and Africa, where about 25% of people live in slums (Gibbs et al., 2010).

Even in developing countries, there is a growing awareness among residents and policymakers about this hazardous position, but it is still insufficient, and achieving sustainability continues to be a major obstacle. Technically speaking, we need to propose community-controlled, small-scale landfills with minimal capital costs and high energy efficiency. Similar recommendations were made by many for ensuring the sustainability of an engineered landfill in developing countries, including the use of simple, easily implemented technologies, low-cost solutions for development, operation, and maintenance, and the best possible exploitation of local natural resources and material properties. The single most expensive component of modern landfills, artificial liner systems, has a negative impact on the project's financial viability. The operational issues with the larger cities' constructed sanitary landfills in DCs include the cost of the daily cover which can account for up to 50% of operating expenses, the management of LFG (landfill gas), and the control and treatment of leachate (Liang et al., 2022). The semi-aerobic landfill stands out among the others as appearing to best meet the needs of developing countries because of its ease of management, low energy demand, and shorter long-term impact as a result of the enhanced aerobic degradation of organics.

A semi-aerobic landfill system is composed structurally of vertical venting pipes built at particular intersections of the horizontal pipes and a network of horizontal pipes inserted at the bottom of landfill sectors. As part of the network of pipelines, the perforated horizontal pipes not only collect and quickly drain off leachate produced in the waste layers but also encourage air circulation within the waste mass. The significant proportion of organic waste that is often found in those countries' trash may provide a problem despite the well-designed ventilation system because of the amount of oxygen required for biodegradation. A natural ventilation system cannot guarantee a completely even dispersion of air throughout the mass, and some methane is created; the higher the percentage of organic matter, the greater the likelihood of an anaerobic situation. It is possible to determine the benefits of using the semi-aerobic landfill in terms of leachate emissions. High putrescible content is a common description of trash/waste in DCs, and anaerobic landfilling is the most popular type of controlled disposal. The concentration of ammonia, volatile fatty acids (VFA), which are a good indicator of the putrescible portion of the waste, and the TOC, which measures the total concentration of organic carbon, are all significantly lower in semi-aerobic conditions than in anaerobic conditions for both types of waste. Similar outcomes were discovered in a number of semi-aerobic landfills, a novel dual-step management method for semi-aerobic landfills created specifically for tropical climates is proposed, alternately recreating a composting phase and a flushing phase during dry and wet seasons, respectively. The collected results showed that the creative management was effective in meeting sustainability target limitations at the conclusion of each year. The horizontal growth of the landfill is necessary for the innovative management, which implies a need for large amounts of area and leads to large amounts of leachate production, which is linked to increased landfill management expenses. To ensure irrigation during the dry season and to improve leachate evaporation and treatability, which reduce quantities and lower management costs, leachate generation is essential if space is not a constraint. A hybrid reactor can be used to manage a semi-aerobic landfill in a wet-dry climate in tropical regions. The hybrid reactor can be aerated during the dry season and flushed in anaerobic conditions during the wet season.

Many solid waste management techniques have been used in low- and middle-income countries, but they frequently fail because of socioeconomic and technical limitations present there(Mor & Ravindra, 2023). Since the 3R-hierarchy and the circular economy approach both require time to be properly applied and, in any event, a step-by-step implementation, enforcing solutions established in the industrialized countries does not guarantee an effective and resilient solid waste management system.

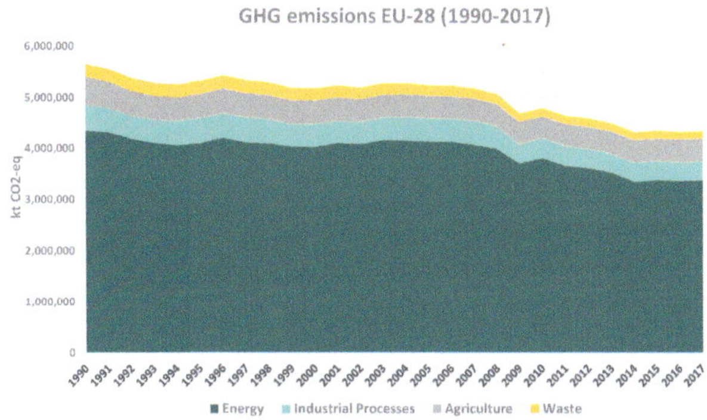

Figure 1.10. Greenhouse gas (GHG) emissions. (Source: González-Sánchez M, Martín-Ortega JL. Greenhouse Gas Emissiong/10.3390/su12031012https://upload.wikimedia.org/wikipedia/en/f/ff/Global_human_greenhouse_gas_emissions%2C_by_sector%2C_2016.png, Creative Commons License).

Correctly quantifying the greenhouse gas (GHG) emissions (Figure 1.10) from the waste sector, improving government capacity for GHG monitoring and reporting, and developing waste management strategies are all necessary. If regionally implemented and appropriately localized, these strategies can contribute to waste diversion. Local authorities in Africa typically lack the expertise to assess technology options, work with limited resources, lack capacity and data, and operate in complex institutional and social contexts, all of which increase the risk of inappropriate technologies failing and installations being made out of context. Furthermore, environmental and socioeconomic factors are typically taken into account secondarily to technical and economic factors while implementing waste reducing plans. Local municipalities have recently been obliged to look into alternative waste management strategies due to the growth of waste management legislation and the needs for landfill development. It becomes necessary to put alternative solutions into practice, such as recycling, recovering biogenic waste, and using trash/waste as a resource. Local governments do not, however, have the necessary human capital or funding to deploy such new systems. Up to 40% of South Africans receive only minimal or no waste/trash services (Rogers, de los Reyes, Beckwith, & Borden, 2013). The Waste Resource Optimization Scenario Evaluation (WROSE) model/tool, which was developed to help municipalities and the private sector evaluate various waste management strategies and make the best and most sustainable decision from an environmental, technical/economic, social, and institutional perspective, aims to close this gap. In order to create a comprehensive waste reduction and climate change stabilization strategy for the waste industry, research was conducted using the Waste Resource Optimization and Scenario Evaluation (WROSE) model to estimate GHG emissions and alternative waste diversion paths from the metropolitan municipalities in South Africa (Friedrich & Trois, 2013). The project aims to give municipal waste managers with statistics and information regarding feasible landfill disposal alternatives, utilizing those alternatives' carbon footprints and potential for GHG reduction as deciding factors.

The WROSE model was created to aid municipal officials in making decisions regarding the implementation of suitable waste management techniques. For municipal officials seeking to adopt alternative waste management strategies through the activation of public-private partnerships, the WROSE model was created in collaboration with the private sector. A variety of waste management technology options are covered by the model, including landfilling, landfilling with gas extraction, landfilling with gas recovery and electricity generation or gas upgrade, recycling, thermal treatment and incineration, anaerobic digestion, and composting. The model uses country-specific data and emission factors, making it applicable to developing countries. The WROSE model also includes basic capital and operating costs for the waste management operations. The model offers a thorough analysis of related capital and operational costs/revenues, prospective job creation and associated health hazards, institutional framework relative to the implementation of the evaluated technology alternatives. WROSE was created with IPCC emissions factors and adheres to industry standards for evaluating carbon emissions based on a first-order decay model. The WROSE technique is thus a relian emission models employed globally, such as WARM, WRATE, EASETECH, or GAINS (Trois et al, Abera, Dell'Orto, Kissoon, & Gómez-Sanabria, 2023)the disposal of unsorted waste to landfill is still the primary waste management method across the country; however, legislative developments aim to drive integrated waste management and the circular economy, putting the disposal of waste to landfill as the least favorable waste management solution. With 80% of the municipal solid waste ending up in landfill sites unsorted and untreated, the only reasonable "activator" of the circular economy would be an integrated waste management system, which is underpinned by an efficient separated collection at the source, followed by carefully selected and strategically localized waste treatment strategies, decentralized recycling facilities, and sustainable end-of-life disposal options. The Waste Resource Optimization and Scenario Evaluation (WROSE.

1.9 GOALS

Typically, the goal of waste management models is to help decision-makers create integrated plans for adopting solid waste management alternatives. Many techniques, including cost-benefit analysis, life cycle assessment (LCA), environmental risk assessment, multi-criteria decision-making, and environmental impact assessment, constitute the foundation of the bulk of these decision support models. The effectiveness or sustainability of waste management is influenced by a number of characteristics, including the quantity and composition of garbage as well as socioeconomic, technological, topographical, and other elements. Through energy and materials recovery, choosing the right waste processing technologies and waste management practices can maximize net energy production, lower costs, enhance waste diversion from landfills, lower GHG emissions, and minimize negative eironmental effects. To estimate garbage generation and disposal in metropolitan municipalities, various statistics have been gathered.

The first stage in gathering trash data was to review the integrated waste management plans (IWMP) (Figure 1.11) and integrated development plans (IDP) for the metro and province. Population, income level, GDP, quantity of waste dumped in landfills, waste composition, waste collection rate, and landfill locations were only a few of the statistics that were gathered. The data study showed that waste categories are determined differently in each of the metro areas. Based on the literature that was available, the following was made: mixed MSW, food waste, garden waste, mixed paper, mixed glass, mixed metals, LDPE, HDPE, PET, and others. It was challenging to anticipate current waste generation patterns due to gaps in the available waste data, inconsistent waste data collection and reporting practices in the IWMPs, and out-of-date waste characterizations for some towns.

The carbon emissions production or reduction potentials in $MTCO_2eq$ were computed using the waste fractions determined above using emission factors from the IPCC (2006). The emissions factor for biologically treating biogenic MSW is 1 g CH_4/kg of wet waste. The amount of methane recovered for energy generation is anticipated to be 95 percent, and very less nitrous oxide emissions. From the time the waste is generated by the waste generator until it is disposed of, treated, or repurposed into new goods, GHG impacts are taken into account. Municipal waste management (MSW) policies are heavily focused on landfilling as the final stage of disposal in Latin American nations. Other significant SWM alternatives, like waste treatment and material or energy recovery, are still unavailable or have only been tried on a small scale. Over the pas have been made to change ultimate disposal techniques from open dumps to designed landfills (Al-Maaded et al, Madi, Kahraman, Hodzic, & Ozerkan, 2012). The percentage of the people that dispose of their solid waste in landfills has considerably increased in countries like Colombia and Brazil.

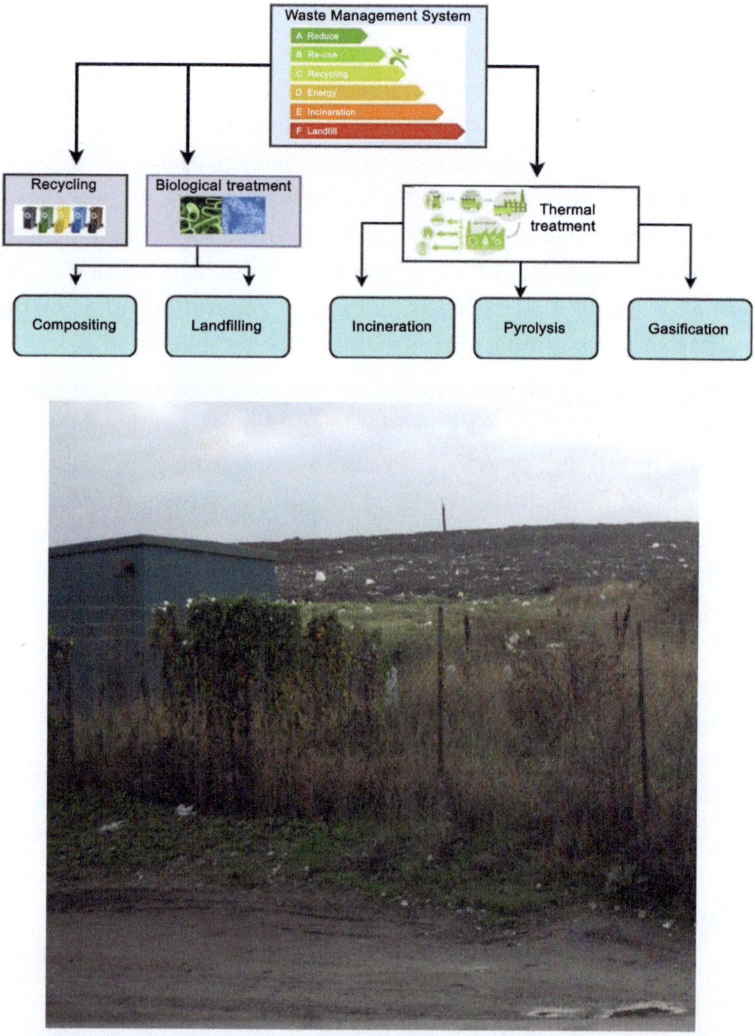

Figure 1.11. Integrated waste management. (Source: Ławińska O, Korombel A, Zajemsk.3390/en15020510https://upload.wikimedia.org/wikipedia/commons/d/db/Rainham_Integrated_Waste_Management_Site_-_geograph.org.uk_-_1601683.jpg, Creative Commons License).

This transition from open dumps to designed landfills may be seen as a positive step in minimizing any potential environmental effects related to the production of MSW. In fact, it is possible to view the final disposal of MSW as a crucial stage in solid waste management because it is during this stage that the greatest environmental effects, particularly those with long-term implications on sensible habitats, can arise. Source separation, recollection, transport, treatment, material recovery, and energy recovery are just a few of the processes that come before final disposal and have an immediate impact on this conclusive stage. In other words, whatever is done in the earlier processes has an impact on the potential environmental effects that MSW final disposal could have .

Landfilling is by far the most popular technological approach for the ultimate disposal of MSW in Latin America. Massive efforts have been undertaken in many countries to switch from open dump methods to constructed landfills, although treatment and mass and energy recovery from waste are still in their early stages of implementation. As was already established, this fact means that the associated environmental damage potential is unquestionably large given that MSW enters landfills without any form of prior treatment or recovery. Leachate production and treatment are primarily responsible for the majority of this possible environmental impact, including long-term issues. The treatment of leachate then becomes a crucial concern in Latin American solid waste management strategies. If the environmental leachate treatment requirements are not met, the transition from open landfills to sanitary landfills will just have changed the problem's nature from a dispersed pollution problem to an end-of-tube one. Furthermore, it must be stressed that a landfill without adequate leachate treatment may have worse long-ronmental effects than leachate from open dumps. Leachate treatment, and more specifically the expenses associated with leachate treatment, istherefore, therefore, one of the primary issues in MSW management in Latin America. MSW management businesses, environmental authorities and regulators, and state bodies are prioritizing the increase in MSW service tariff that can result from implementing leachate trent. The key is to figure out how to adhere to cost-effective requirements while yet meeting environmental standards for leachate discharge quality.

CostThe cost-effectiveness of leachate treatment relies on treatment goals. It is still unclear how environmental authorities in Latin American nations decide on environmental criteria, including, in this case, leachate treatment goals. In most cases, regional and local specificities are not taken into account while establishing environmental guidelines or permitted discharge limits. In this regard, the pollutant assimilative ability of water-receiving bodies is a fundamental factor that environmental authorities sometimes overlook when defining leachate treatment objectives. Obviously, this could result in expensive leactnt.

On the one hand, it's possible that the treatment targets established do not take into account the capacity of the receiving water body's capacity to absorb pollution. As a result, the contaminant removal needs may be excessive, which would result in cost inefficiencies. Even if the effluent quality meets the objectives imposed, the environmental impact cannot be avoided, which would result in inefficient costs. Research studies examined the cost-effectiveness of leachate treatment in Latin America with the goal of establishing a conceptual framework for comprehending the notion of cost efficiency applied to leachate treatment and demonstrating some study cases in which some leachate treatment expediencies in Latin America are evaluated from a cost-efficiency approach. The effectiveness of leachate treatment costs depends, first, on attaining the suggested treatment objectives and, second, on incurring the lowest possible overall cost to do so. Therefore, effective leachate treatment costs are those linked to technologiable achieving treatment goals with the least amount of capital and operational expenditure. Then it is suggested that state regulatory agencies recognise recognize effective costs in the MSW tariff structure.

Everything that is done to manage MSW has an impact on any potential environmental effects related to the final disposal. Since the most substantial environmental impacts, particularly those involving long-term effects on sensible habitats, can arise during the final disposal of MSW, this phase of solid waste management can be viewed as crucial. All the procedures that come before ultimate disposal, like source separation, collection, transport, treatment, and mass and energy recovery, have an immediate impact on this stage, which is the last and most important in terms of sustainability. In other words, whatever is done in the earlier processes has an impact on any potential environmental effects that landfilling might have. The sole technological method of MSW management available in the majority of cities in Latin America is landfilling (Ferronato et al., 2023). It denotes the absence of any sort of treatment or recovery prior to garbage disposal. In Latin America, the organic matter fraction in MSW is particularly significant. More than 60% of the total MSW produced in an urban region is referred to as biowaste. The production of biogas and leachate is a significant concern with regard to the origination of environmental impacts because biowaste is by far the largest MSW fraction that is landfilled. Leachate can have an impact on water resources, soils, and potential health hazards whereas biogas may produce greenhouse gas emissions.

Figure 1.12. Biowaste. (Source: https://upload.wikimedia.org/wikipedia/commons/a/af/Foreign_objects_separated_from_residential_biowaste.jpg, Creative Commons License).

When biowaste (Figure 2.12) makes up the majority of the waste mass disposed of, the leachate composition is strongly characterized by high organic matter and nitrogen concentrations. Latin America is currently transitioning from open dumps to landfills, which is undoubtedly a development in MSW management. However, it poses a problem for solid waste service providers and environmental agencies in terms of handling leachate treatment and discharges. Operators of solid waste services must meet the technical and financial requirements, and environmental authorities must be able to set reasonable leachate treatment standards and oversee the functioning of the entire landfill.

1.10 ENVIRONMENTAL RISK

The size and geographic scope of environmental effects related to landfilling are primarily influenced by the complex composition of leachate. It depends on the following elements: local meteorological conditions i.e., precipitation, temperature, and evaporation; the kind and operation features of the landfill; the physical, chemical, and microbiological properties of the confined waste; and the age of the confined waste. Four categories of chemicals can be found in leachate: dissolved organic matter, inorganic macro-components and heavy metals, xenobiotic organic compounds, and other substances. Leachate has an impact on the environment when it comes in contact with the ground or a surface water source, changing some or other quality attributes of these environmental matrices. If a receiver i.e., a population or community is exposed to the compounds, these envirtal effects may even offer a health risk concern. The effects on water sources exposed to organic loads are referred to as the environmental effects related with to the presence of dissolved organic matter (BOD or COD) (Hansen et al., 2016)little information is available to assess the impact of biological and photolytic processing on the optical properties of original DOM source materials. Over a 3.5 month laboratory study, we measured changes in commonly used optical properties and indices in DOM leached from peat soil, plants, and algae following biological and photochemical degradation to determine whether they provide unique signatures that can be linked to original DOM source. Changes in individual optical parameters varied by source material and process, with biodegradation and photodegradation often causing values to shift in opposite directions. Although values for different source materials frequently overlapped, multivariate statistical analyses showed that unique optical signatures could be linked to original DOM source material, with 17 optical properties determined by discriminant analysis to be significant ($p < 0.05$. The reduction in dissolved oxygen brought on by the aerobic biological oxidation of the biodegradable portion is the well-known environmental effect. High BOD readings suggest that this is likely to have an effect on a superficial water supply, such as those that occur during the acid phase of landfill leachate operation. The potential for leachate to have this impact, however, is significantly diminished as the ratio of BOD/COD approaches zero, which is the normal situation of mature leachate (methanogenic phase), which is for uncertain quantities of BOD.

Nitrogen is without a doubt the inorganic macro-component with the biggest impact on human health and the environment. Given the anaerobic conditions present in a landfill, ammonia nitrogen ($N\text{-}NH_4^+$) is the major form of nitrogen, and there are no differences in concentrations between the acid phase and methanogenic phase. This indicates that ammonia nitrogen concentrations do not diminish during the methanogenic phase or even during the period following the closure of a landfill, making this chemical the primary problem given its enormous potential influence on water supplies. Because of the potential risk to public health that heavy metals pose, their presence in leachate has historically attracted a lot of attention. But generally speaking, the levels of heavy metals in leachate that have been documented in the literature are minimal. The highest concentrations of these substances are found in the acid phase due to lower pH values that increase the solubility of the metals, whereas in the methanogenic phase, neutral pH favors the phenomena of adsorption and precipitation within landfills, and as a result, low concentrations of heavy metals in mature leachate are found. The term xenobiotic organic chemicals refer to substances like insecticides, phenols, aliphatic chlorinated compounds, aromatic hydrocarbons, and halogenated hydrocarbons. Due to the detrimental effects many of these chemicals may have on aquatic environments as a result of their potential toxicity, several of them are regarded as priorities. Fresh leachate concentrations for these chemicals, however, are typically low and have been recorded in the literature. It is anticipated that over time, biological breakdown mechanisms will lead to lower amounts of xenobiotic organic compounds.

The expenses involved in treating leachate are directly related to the requirements for meeting treatment goals or effluent regulations. While the goal of establishing leachate effluent regulations is to safeguard receiving water bodies from dangerous pollutant concentrations, the process of establishing leachate treatment goals differs widely from country to country. The quantity and quality of leachate produced, as well as the treatment goals set by the relevant environmental authority to meet effluent regulations, are thus the factors that affect the costs of leachate treatment. It has been noted that treatment goals must be established in terms of the required quality of the effluent in order to account for the potential environmental effects that leachate may have depending on its composition. The receiving water body is a crucial component of the local characteristics. It is then necessary to conduct environmental studies that consider ongoing monitoring of the quality of the receiving water body and, more importantly, simulation of the transport and fate of contaminants present in leachate in order to set leachate treatment objectives that are consistent with an analytical procedure. In light of the water quality requirements for the source's various uses, it is conceivable to estimate the assimilative capacity of the water-receiving body from these research studies. The objectives of leachate treatment can therefore be determined, as well as how they must be modified over time in accordance with the anticipated changes in leachate production, both in quantity and quality.

Due to a lack of environmental data regarding water bodies and basins and the high costs of environmental studies comprising simulation of pollutant movement and fate, the analytical approach is not frequently used in some countries. Unfortunately, leachate treatment goals are frequently established at random, possibly in accordance with some international effluent discharge standards (Nehrenheim, Waara, & Johansson Westholm, 2008). This circumstance has the potential to significantly increase the cost of treating leachate. The effectiveness of leachate treatment depends on two factors: first, achieving treatment goals for the effluent discharge quality; and second, fulfilling the least cost criterion linked with the use of technologies that may achieve the corresponding pollutant removal. Therefore, it may be claimed that the costs of effective leachate treatment are those incurred while the treatment system achieves the effluent discharge quality at the lowest possible cost. Therefore, national regulatory tariff strategies should be focused on achieving effective leachate treatment costs. Two circumstances in this context could make leachate treatment inefficient in terms of cost. First off, it's possible that the amount of pollutant removal required by the treatment system is very large when the treatment objectives are arbitrarily established without taking the assimilative capacity of the receiving water body into account. When organic matter is the harmful pollutant and a large-flow river is the receiving water body, the first scenario may happen. Second, when the treatment goals are arbitrarily defined and fail to take into account the receiving water body's poor assimilative capacity. This indicates that substantial environmental damage to the water source is still possible even if the effluent discharge quality is fulfilled via leachate treatment (Jagaba et al., 2021)due to the risk of leachate infiltration into to land and aquifers. The generated leachate is considered a serious environmental threat for the public health, because of the toxic and recalcitrant nature of its constituents. Thus, it must be collected and appropriately treated before being discharged into the environment. At present, there is no single unit process available for proper leachate treatment as conventional wastewater treatment processes cannot achieve a satisfactory level for degrading toxic substances present. Therefore, there is a growing interest in examination of different leachate treatment processes for maximum operational flexibility. Based on leachate characteristics, discharge requirements, technical possibilities, regulatory requirements and financial considerations, several techniques have been applied for its degradation, presenting varying degrees of efficiency. Therefore, this article presents a comprehensive review of existing research articles on the pros and cons of various leachate degradation methods. In line with environmental sustainability, the article stressed on the applicationd efficiency of sequencing batch reactor (SBR.

The identification of alternative applications to extract resources along the pipeline before disposing of materials to in a landfill can change the flow of materials from a linear flow or disposal to a circular flow, which can be realized as one potential driver of sustainable development. Collective efforts fom the public and business sectors are guiding the transition to a circular economy. Municipal waste management (MWM) is regarded globally as a public good like others i.e., transportation, water, and power distribution that necessitates adherence to a set of laws and sound governance amongst the various parties involved. In the majority of countries, local governments are responsible for providing this service as one of their municipal responsibilities; these responsibilities are often governed by a national legal framework. In industrialized countries, national governments enact rules and regulations that specify goals, benchmarks, and operational and environmental standards. Effective and efficient solid waste management (SWM) techniques support resilient city governance and planning. SWM in underdeveloped countries is frequently characterized by a lack of regulation and poor governance. Where laws are in place, they are typically not strictly enforced. A significant part of the system is played by the informal sector, which makes governance and law enforcement more challenging. There are still reported governance problems in many other developing countries, which are said to account for their low SWM management efficiency.

Under the combined impact of rapid urban growth and economic development, waste management in MENA countries will pose a significant problem in the coming years. Governments in the majority of them are aware of how serious the issue is and are working hard to put in place a strong legal system. Despite these efforts, the implementation of regulations is still in progress and only represents one side of the equation. The other side of this coin is its enforcement, which is considerably more difficult because it calls for a solid governance structure that enables effective oversight and management. The informal sector plays a role in recycling in underdeveloped countries since products like glass, plastic, paper, and metal have created markets. Appropriate steps should be made to assist these initiatives, including increasing public awareness, adopting particular laws and regulations, and putting in place sustainable waste management infrastructures. Political, institutional, environmental, social, and economic factors all have a role in the complexity of the SWM. Building sustainable municipal SWM that satisfies public demands might be difficult for local governments in poor countries due to a fundamental cost constraint. Environmentally responsible waste management can be encouraged and the harmful impacts of landfilling and incineration can be avoided with the aid of a political framework and policy.

CHAPTER 2

Problems with Waste Management

CONTENTS

2.1 INTRODUCTION

One effect of global urbanization (Figure 2.1) is an increase in waste production and a variety of characteristics. In low-income countries, more than 90% of the waste produced is typically burned or dumped in open areas(Ferronato & Torretta, 2019). Uncontrolled waste generation and poor waste management have resulted in a variety of environmental and health problems. Solid waste management is an important administrative burden for both small communities and large cities.

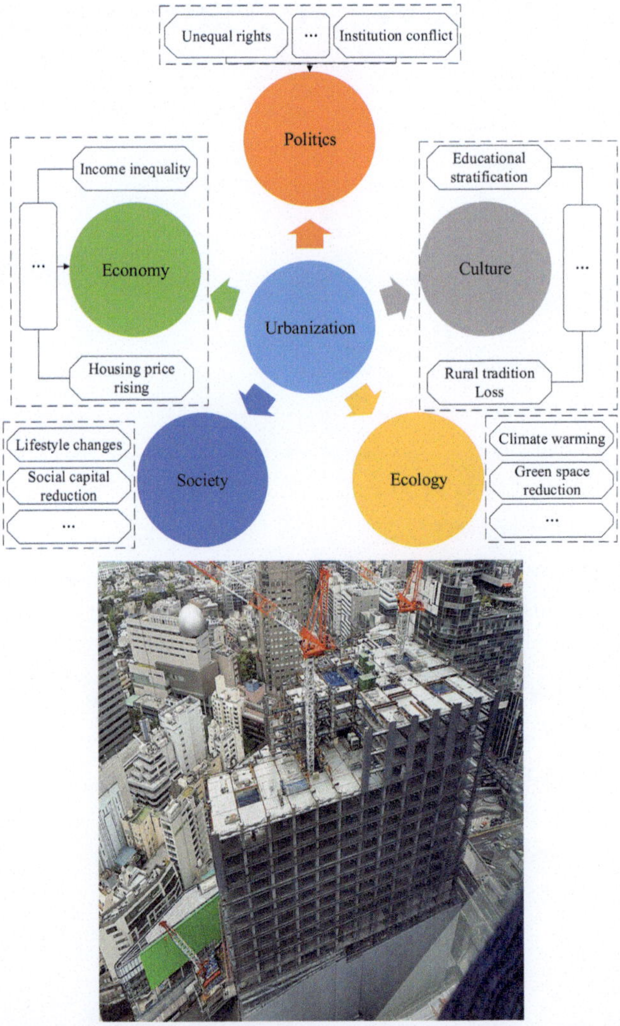

Figure 2.1. Urbanization. (Source: Li C, Yan J, Xu Z. How Does New-Type Urbanization Affect the Subjective Well-Being of Urban and Rural

Residents? Evidence from 28 Provinces of China. *Sustainability*. 2021; 13(23):13098. https://doi.org/10.3390/su132313jpg, Creative Commons License).

A long-term effective and successful waste management policy requires an understanding of waste characteristics. It is the first stage in determining the amount of material that might be recovered, locating the sources of waste production, facilitating the design of processing machinery, analyzing the physicochemical characteristics of the wastes, and maintaining compliance with regulations. Seasonal variations, human lifestyle, population dynamics, geography, and local regulations all affect the quality and quantitative features of waste that is produced. Over time, the definition of solid waste evolves importantly. Solid waste is any waste that consists of solids, non-soluble materials, semisolids, including gases and liquids in containers, and that is generated by domestic, mining, agricultural, industrial, commercial, and other activities. It is crucial to remember that solid waste is not just physical solid. Some solid wastes are found as liquids, while others are semi solids or gases. Some items that were once discarded such as radioactive waste, home sewage, point source discharge, in-situ mining, hazardous secondary waste from the petroleum sector, and coke products have been removed from the category of solid waste in the twenty-first century.

A society's status is influenced by its surroundings, cultural values, historical context, and economic state. A clear understanding of the state of society offers guidance for addressing the waste problems. As civilization develops over time and becomes more affluent and advanced, it becomes to rely on the extraction and metabolism of massive amounts of resources, especially energy, in order to sustain the enduring population. The high level of solid waste production was a natural result of increased resource use. All processes involve the extraction of raw materials, the production of finished goods, and the consumption of materials to produce solid waste. As far as we are aware, according to the principles of thermodynamics, materials and energy can only be transformed from one form to another. Because the wastes of one organism are used as food by another, there is virtually no waste produced.

This organic method of material recycling adheres to the sustainability principle. There are 1.7–1.9 billion metric tonnes of solid waste generated worldwide (Rajaeifar et al., 2017)developed countries have implemented sustainable material management (SMM. By the middle of the twenty-first century, it is predicted that there would be 3.40 billion tonnes of solid waste produced. This amount is greater than the population growth rate, which over the same period of time doubled. Clearly, the amount of waste produced and its physical and chemical qualities can vary by country.

2.2 WASTE

Waste production in wealthy countries may increase by 19% in 2050 compared to developing countries, where it is anticipated to increase by between 40 and 50%(van Puijenbroek, Beusen, & Bouwman, 2019). The continents with the fastest population growth are Africa and South Asia, where in 2050, total waste production would more than triple and double, respectively. The majority of solid waste is openly dumped in these areas, and the manner in which it is produced has important negative effects on health and economy. The control of waste generation presents difficulties for Nigeria, the most populated country in Africa. The city of Lagos, the nation's quickly developing capital, is referred to as the waste capital of the world. This city's residents regularly dispose of 11,000 tonnes of solid waste. Meanwhile, the city struggles to generate electricity; at times, they can only manage 3–4 hours per day. The development of technology, acceptance of waste reduction strategies, and study in material sciences may decrease waste generation in industrialized countries in the world. On the other hand, as we all know, poor countries lack the financial resources, effective leadership, and opportunities for research and development. The amount and nature of waste generation in rich and developing countries may also differ for this reason. We need to change because of the growing global problem of solid waste and its severe social and environmental efcts.

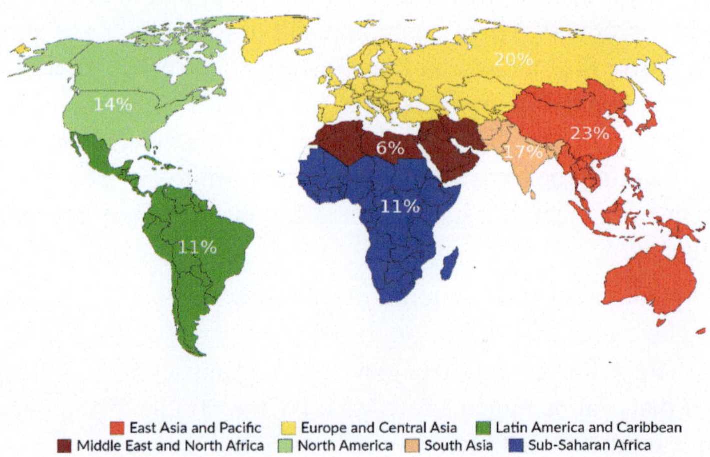

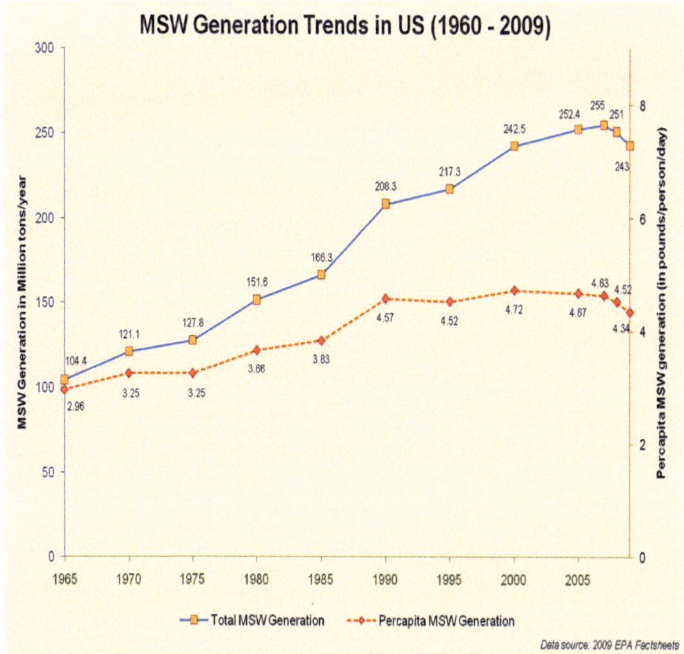

Figure 2.2. Solid waste generation. (Source: Namoun A, Tufail A, Khan MY, Alrehaili A, Syed TA, BenRhouma O. Solid Waste Generation and Disposal Using Machine Learning Approaches: A Survey of Solutions and Challenges. *Sustainability*. 2022; 14(20):13578. https://doi.org/10.3390/surends.jpg, Creative Commons License).

Solid waste generation (Figure 2.2) is a global phenomenon that is particularly prevalent in urbanized regions. SWG is regarded as the most difficult problem faced by developing countries, which experience severe environmental risks as a result of high SWG. High solid waste generation in metropolitan areas sparked hygienic problems with infrastructure, water supplies, and waste management. Inadequate waste management rules, such as waste collection services, municipal and national agencies to handle the solid wastes, are present in developing countries. The first is a lack of genuine stakeholder involvement in decision-making and planning procedures. Because understanding the changing relationships among governmental authorities, decision-makers, and societal dynamics depends on their involvement. Because it is a challenging process, many different parties must work together. For instance, government organizations have not worked with municipal organizations to choose appropriate sites for waste management. Municipalities face a problem because they lack an effective organizational structure and qualified personnel. Campaigns to promote waste management are not being led by politicians.

2.3 CHALLENGES

In developing countries, NGOs and private institutions have enough funding to address solid waste challenges. There is a lack of understanding that the system is made up of many hierarchies, including generation points, waste collecting locations, transit, transfer, and final disposal sites. We need to acknowledge that the system is influenced by outside forces, including monetary, institutional, environmental, technical, social, and legal. Poor knowledge, insufficient data, and challenges collecting accurate estimates of waste quantitative analysis are characteristics of developing countries. There are many causes, including inadequate finance, a lack of management expertise, pressing problems to be resolved, and ineffective local government. Due to a lack of open data sources, improper waste management, and an influx of rural-urban migration at the national level, getting reliable data from developing countries is typically challenging. The Philippines, a developing nation, is regarded as Southeast Asia's third-largest producer of solid waste. By 2030, every person will produce five kilograms more waste on average (Oenema et al., ilippines' metropolis, Manila (Philippines), is leading by the worst exang emergingdeveloping countries. An estimated 8000 tonnes of waste are produced every day by its residents. Unfortunately, despite being an important stakeholder, the government was unable to collect the waste or educate the populace about waste reduction methods. As a result, there were waste piles at numerous dumps, which served as a habitat for both animals and insects. The dumps also encouraged those in need to collect goods from waste to make money to support themselves. The poorest residents even resided atop the landfill that released toxic odors and methane. 219 persons lost their lives in the Payatas typhoon disaster that occurred two decades ago (Wisner & Uitto, 2009). After the tragedy, the government cleaned up the area, but regrettably, a new landfill opened up, continuing to be a source of income for many scavengers.

Sweden has set a good example for wealthy countries when it comes to managing solid waste since they make good use of waste for local development. About 99 percent of all domestic waste in Sweden is recycled or reused, and it is then turned into energy and other products. For instance, biogas is utilized for transportation and the production of power, as well as for heating homes during the winter. Their success can be attributed to the great public participation. They are teaching kids the need for recycling from an early age. In many places, the manufacturing firms are in charge of treating the waste generated by their goods in an appropriate manner. The Swedish government pushes manufacturers to create more recyclable, efficient products. Sweden, a country that is a part of the EU, places a high priority on protecting the environment and has strong laws governing resource management in the food industry. As part of an effort to minimize food waste by 50% by 2030, stakeholders in the food chain business have signed an agreement. This is in accordance with the Sustainable Development Goals. Sweden has been hailed as the world leader in waste management. Based on its composition and sources, solid waste is divided into different categories. Plastic, paper, rubber, leather, culinary leftovers, glass, metal, and garden waste are all components of solid waste. Markets, municipal facilities, factories, agricultural operations, and demolition sites are among the sources of the waste. Industrial waste is frequently subject to environmental regulations depending on its hazardous characteristics, such as ignitability, corrosivity, and toxicity.

2.4 WASTE COMPOSITION

The physical presence of various sorts of components in solid waste is referred to as the composition of waste. It is typically found by carrying out a routine field survey in the desired locations. Waste samples are taken from producers or final disposal sites and divided according to content. High-income or developed countries produce comparatively little organic waste, which accounts for only 32% of total waste, as well as an important amount of dry waste that can be recycled, which accounts for roughly 51% of total waste. As the GDP level declines, developing countries create 53% of the food waste and 56% of the green waste (Lee, Ng, Kwong, & Tay, 2019). Without suitable procedures, developing countries cannot control the continuously changing waste composition. These compositional variances shed light on the purchasing habits, style of living, and financial situation ofin emergingdeveloping countries. Only organic waste, which accounts for 74 g of India's daily total waste production, is produced at a rate of roughly 91.01 g per person. A recent study found that the average household waste output had a negative link with the number of family members and a positive correlation with household financial status and education level. Nearly 82 percent of the waste is organic, which has a high recovery rate. The annual amount of total organic waste is around 232 Gg.

Municipal olid waste is the collective term for materials that are wasted and come from both residential and commercial settings, but not from industries. Examples include cardboard, food wastes, steel, iron, pet bottles, yard wastes, kitchen wastes, plastics, metals, glass, and E-waste. According to their origins, such as urban and peri-urban areas, solid waste in the municipality is made up of a variety of uniform wastes. Due to the swelling demand for goods and services, the pace of MSW generation is higher than the growth in the urban population. In affluent countries, the majority of MSW is disposed of in landfills, and some hazardous waste is incinerated to destroy viruses. According to a number of studies, families in poor countries produce the majority of municipal solid waste, followed by workplaces or commercial areas. A large portion of it winds up in open landfills, where scavengers find the stuff, they can recycle and sell to make money. A portion of waste that has an economic benefit and makes up 54 percent of all waste has a moisture level of 60 percent. This percentage of waste is comparable to that found in several other countries, including Nigeria, the Philippines, India, and Nigeria. Pakistan experiences a 2.4 percent annual population growth. Pakistan, one of the ten most populous countries in the world, is currently home to 212.2 million people. The capital of Punjab, Lahore, is regarded as Pakistan's second largest metropolis. Through an urbanization movement for the improvement of standard lifestyle and economy, this city has been upgraded. With 0.65 kg per person, the daily SWG would be 7150 tonnes. MSW is made up of biodegradable, paper, plastic, textile, and diapematerials.

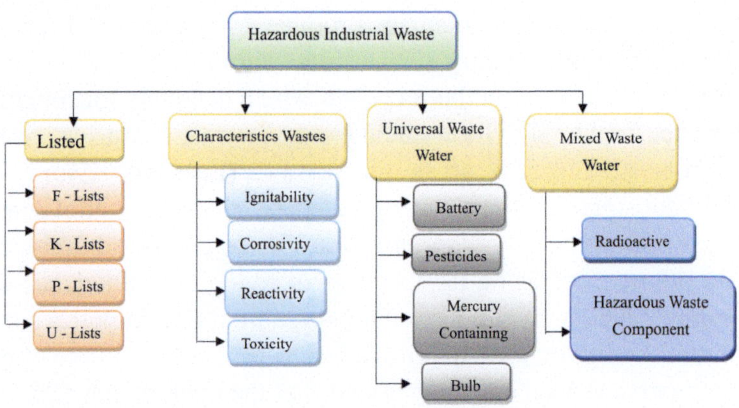

Figure 2.3. Industrial waste. (Source: Palani G, Arputhalatha A, Kannan K, Lakkaboyana SK, Hanafiah MM, Kumar V, Marella RK. Current Trends in the Application of Nanomaterials for the Removal of Pollutants from Industrial Wastewater Treatment—A Review. *Molecules*. 2021; 26(9):2799. https://doi.org/10.3390/molcollection.jpg, Creative Commons License).

2.5 INDUSTRIAL WASTE

Material from mines, factories, chicken farms, construction materials, manufacturing processes, and the provision of goods and services all make up industrial waste (Figure 2.3). This form of waste, which includes soil, oil, chemicals, concrete, and gravel, may be solid, liquid, or gaseous. Non-hazardous and hazardous waste are the two types of industrial waste. Non-hazardous industrial wastes, such as Carton, Plastic, Metals, Glass, Rock, and Organic Waste, are characterized as having qualities between municipal and hazardous waste, whh do not pose a harm to human health or the environment. Because hazardous industrial waste has the potential to be dangerous if it is handled improperly, this risk depends on the concentration, chemical, or infectious characteristics of the waste, which may raise the risk of disease, death, or environmental damage. Oranic substances and non-degradable toxic metals are the two major categories of hazardous waste. Ignitability, reactivity, toxicity, and other four chemical properties are utilized to identify substances that may be harmful. This type of waste ignites easily and is flammable. Paints with solvent bases, petroleum compounds like gasoline, detergents, and other waste materials, for instance, are all flammable. The majority of them come in both liquid and solid forms. lid waste has a 's propensity for ignition, which makes it potentially dangerous.

These pollutants, such as hydrocarbons, electroplating waste, sulfuric acid, oil, and pesticides, are acidic or basic in nature and exist in aqueous solutions with pH levels of 2.0 or higher. Under typical conditions, reactive wastes, such as lithium, sulfur batteries, and explosives, go through violent chemical reactions. When this reactive waste reacts with water at high temperatures or under ideal circumstances, it can be toxic and explosive. It is the capacity of a chemical to induce harmful consequences in a live organism upon exposure. Industrial waste contains toxic chemicals or compounds that leak into water tables and reduce the quality of drinking water. These substances include mercury, cadmium, chmium, and arsenic.

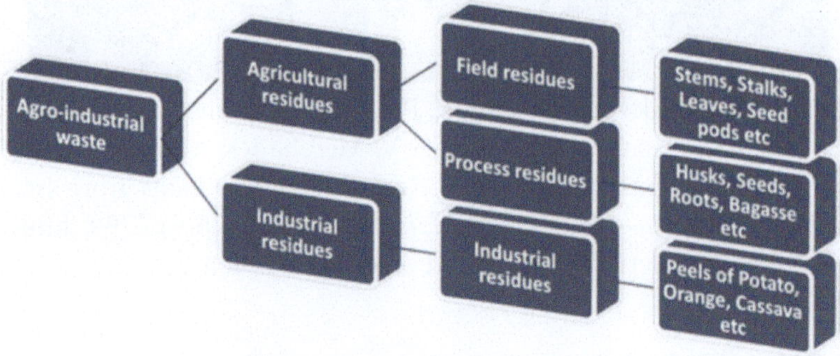

Figure 2.4. Agricultural waste. (Source: Khan HM, Iqbal T, Yasin S, Ali CH, Abbas MM, Jamil MA, Hussain A, M. Soudagar ME, Rahman MM. Application of Agricultural Waste as Heterogeneous Catalysts for Biodiesel Production. *Catalysts*. 2021; 11(10):1215. https://doi.oaph.org.uk_-_1240814.jpg, Creative Commons License).

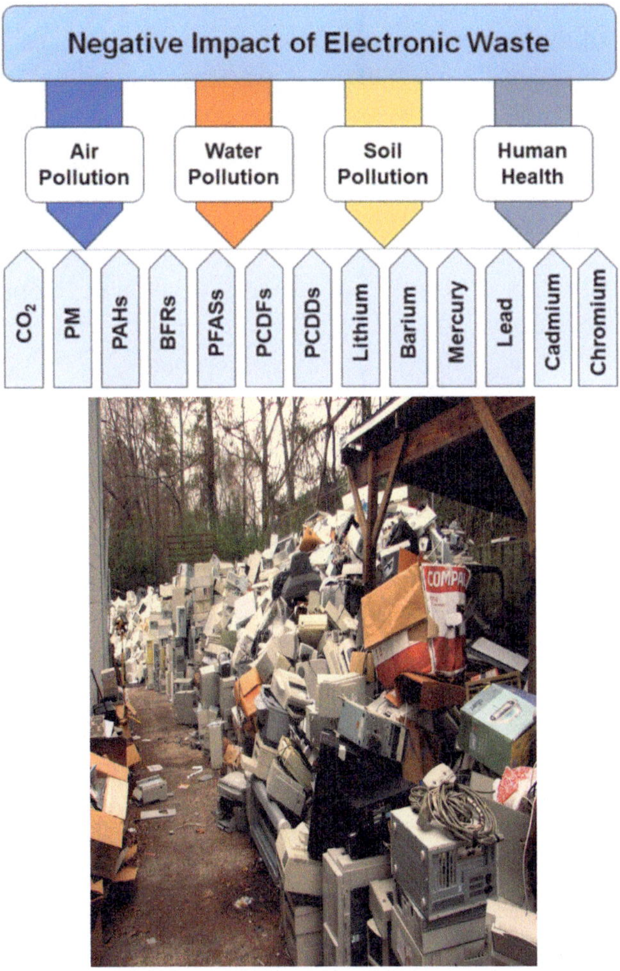

Figure 2.5. E-waste. (Source: Ghulam ST, Abushammala H. Challenges and Opportunities in the Management of Electronic Waste and Its Impact on Human Health and Environment. *Sustainability*. 2023; 15(3):1837. https:ns/6/65/Electronic_waste.jpg, Creative Commons License).

Agricultural methods are the main cause of agricultural waste (Figure 2.4) products such as bagasse, corn, cotton, rice, and wheat straw (Koul, Yakoob, & Shah, 2022). These byproducts contain a lot of calories. These waste materials include chemical elements including ash, carbon, oxygen, and nitrogen. However, waste that is mostly generated by vegetable packing companies does not fall under the category of agricultural waste. Medical waste generation is thought to be primarily fueled by biological products used in human and animal therapy and diagnostic procedures. Laboratory samples, media cultures, lancets, syringes, body parts, and fluids are the key components of these items. Human health may suffer as a result of medical waste. However, these are not the only waste items; animal waste, human anatomical waste, and waste from microbiology and biotechnology are all included in the cegory of medical waste.

2.6 E-WASTE

E-waste (Figure 2.5) is disposed of casually in underdeveloped countries, which has negative effects on human health and environmental damage(Masud et al., 2019). On the other hand, workers in wealthy countries face important risks when recycling and processing electronic debris. To manage risks during recycling activities, the appropriate safety precautions should be used. Tetrabromo-bisphenol A, chromium VI, PVC, lead lithium, mercury, toner dust, and other radioactive substances are examples of e-wastes. The aforementioned chemical compounds must be handled before disposal. These E-waste compounds contribute to ailments like cancer, pneumonia, liver, and kidney damage. Humans may be more susceptible to environmental risks brought on by these substances, such as soil erosion and water contamination. Numerous factors, including financial level, educational attainment, population density, and level of human development, can influence both the quality and amount of solid waste. The need for housing, food, and other natural resources is being put under tremendous pressure by the population growth. SWG has been importantly increased by the rise in community lifestyle standards, growing population, and urbanization. The rapid population growth in the majority of developing countries has made it difficult for the different governments to provide their residents with a decent and quality of life, as required by their constitutions. Recent statistics show that low-income countries are home to 83% of the world's population. Urban populations are growing rapidly in parts of Africa and Asia. This sharp rise in urban population ultimately results in a sharp rise in SWG.

Human behavior is the main cause of the rising SWG, and modifying that behavior is the only way to solve the problem. Societal perceptions and attitudes about waste might have an impact on the entire SWM process. Researchers have also discovered an increase in SWG as a result of the rise in global food demand and other demands, which led to a rise in the amount of waste produced daily by each family member. Some of the most important factors influencing the characteristics of household waste include consumption and habitation patterns in certain regions and during particular seasons. Summertime is typically when more stable waste is produced, which has led to apparent changes in waste properties. People with low levels of education are expected to dispose of their waste on the street and other undesignated areas provided by the local governments. People with little education are against waste collection and sorting services. The key elements in lowering waste generation are public engagement and awareness. The public's participation is an essential instrument for creating a more wholesome and clean society. Without public involvement, resource use will be less effective, which will have a negative impact on sustainability. Educating the public about waste separation can help create an effective waste management system and cut the cost of disposing of it. Sociodemographic and economic factors, including average family size and structure, job status, and monthly income, have an impact on SWG. Social activities and waste composition in a given society are directly related to one another and have an impact on one another. Research was carried out in Indonesian cities. It was revealed that the city had high levels of waste generation, high levels of population density, and high rates of economic growth. There are numerous environmental problems that are fundamental to society and in which households play an important role. For instance, a household with children produces extra waste made up of diapers. Households with only older residents produce less domestic waste than those with a mix of ages.

Ethiopia, one of Africa's rising countries, faces numerous environmental difficulties as a result of economic activity and rapid population surge (Cohen, 2006). The amount of waste produced by each household was positively correlated with family size. Education level, on the other hand, showed a negative association. The same findings were reported in Bangladesh and Nigeria by other investigations. The impact of education on people's attitudes toward waste generation may be a factor in this. Additionally, researchers discovered a positive relationship between SWG and socioeconomic status, level of education, and cultural practices. Due to cultural differences, there is no discernible correlation between the income of each family member and the rate of waste generation. The average amount of solid waste produced daily in Jimma, Ethiopia, was 0.55 kg, which is comparable to Malaysia's average per-capita production of 0.5 kg of waste (Getahun et al., 2012). Almost 87 percent and 13 percent, respectively, of waste production was home and institutional. The occupation of agriculture is practiced by almost 70% of the people. Rapid population expansion in developing countries has given rise to a variety of infrastructure problems and problems with land use planning that are beyond the capacity of national and local governments. Unfortunately, the low level of technology development and infrastructure growth increases the high reliance on raw materials and restricts the use of waste reduction strategies like recycling. Poor landfill infrastructure exacerbates SWG management.

It is a difficulty because waste is frequently moved from higher-income to lower-income locations where there are fewer environmental restrictions and less protection for workers who work with waste, both within countries and between countries. Seasonal variations in temperature, humidity, and rainfall have an impact on the properties and volume of solid waste. Rainfall and humidity immediately change the water concentration of solid waste. Flooding creates mounds of domestic waste, which increases the load on disposal facilities close by. Higher summer rainfall frequency raises the risk of floods, which has an impact on people's access to medical services, amenities, water quality, and the use of mobile waste disposal facilities. Following glacier thawing, rising sea levels will hasten the erosion of coastal dumping grounds, which will pollute the waters along the coast. In most rural areas of developing countries, coal and wood are typically used to heat homes during the winter. Both of these materials produce ash that is disposed of with household waste and changes the composition of ambient air gases. The living environment has an impact on solid waste, waste composition, and level of awareness. In Indonesia, persons who live in urban areas are slightly more conscious of solid waste management than those who live in suburban areas. Less waste is produced every day in the city center, compared to the suburbs.

It is a difficulty because waste is frequently moved from higher-income to lower-income locations where there are fewer environmental restrictions and less protection for workers who work with waste, both within countries and between countries. Seasonal variations in temperature, humidity, and rainfall have an impact on the properties and volume of solid waste. Rainfall and humidity immediately change the water concentration of solid waste. Flooding creates mounds of domestic waste, which increases the load on disposal facilities close by. Higher summer rainfall frequency raises the risk of floods, which has an impact on people's access to medical services, amenities, water quality, and the use of mobile waste disposal facilities. Following glacier thawing, rising sea levels will hasten the erosion of coastal dumping grounds, which will pollute the waters along the coast. In most rural areas of developing countries, coal and wood are typically used to heat homes during the winter. Both of these materials produce ash that is disposed of with household waste and changes tosition of ambient air gases. Living The living environment has an impact on solid waste, waste composition, and level of awareness. In Indonesia, persons who live in urban areas are slightly more conscious of solid waste management than those who live in suburban areas. Less waste is produced every day in the city center, compared to the suburbs.

Figure 2.6. Tourism and waste. (Source: https://upload.wikimedia.org/wikipedia/commons/d/d6/Garbage_along_road_Tourism_ATR_P1180860.jpg, Creative Commons).

2.7 TOURISM AND WASTE

In emergingdeveloping countries like Kenya, India, Pakistan, and the Philippines, tourism is an industry that is rapidly expanding and is a major contributor to the rise in the generation of MSW (Figure 2.6). The systematic waste produced by the tourism industry is still unaccounted for and concealed by domestic waste flows. Studies found that a typical visitor cn produce more municipal solid waste than local people. The hotel sector produces the majority of waste. The majority of the time, tourist destinations are situated in remote locations without adequate MSW management infrastructure or services. This combination may result in a destructive cycle of tourism that destroys the environment and local populations. For instance, Pakistan's Ayubia National Park, one of the country's most captivating and stunning tourist sites, preserves three distinct forest ecozones: subalpine meadows, moist temperate conifers, and subtropical pine forests. The park receives about 120,000 visitors during the summer. Due to pressure from tourists, hotels and restaurants have been producing 2940-3225 kg of waste every day for the past few years. It is concerning because tourists produce 3.38 to 3.84 kg of solid waste per person per day. Growing solid waste severely depletes water resources and disrupts biodiversity on a continuous basis. Threats to public nd environmental security in emergingdeveloping countries include fast urbanization, excess production, irregular waste collection, inadequate sorting systems, open dumping, burning, unpleasant odors, and incorrect disposal of MSW.

A sizable area has been set aside for a waste disposal facility, which is an opportunity, especially in a nation with little available land. These land regions could potentially be used for environmentally friendly development to boost the GDP. It is urgent to look into the underlying problems contributing to the solid waste buildup and to take action to slow the rate of generation. Additionally, the rising rates of malnutrition, severe humanitarian crises, and starving deaths in many devoping countries necessitate programmes to raise awarness of food waste. Greenhouse gases released by solid ste play a role in climate change (C. Zhang, Guo, Wang, & Chen, 2019)limited-controlled landfill is a common method used to dispose of domestic solid waste, especially in developing countries. However, information about GHG emissions from limited-controlled landfill sites has rarely been reported. In this study, the GHG emissions from a typical limited-controlled landfill site were investigated under a regular period for one year. The number and positions of static chambers were arranged according to the guidance on Monitoring Landfill Gas Surface Emissions by the UK Environment Agency to obtain representative data from the heterogeneous surface of the landfill. Inverse distance weighting (IDW. Low-income economies account for the majority of GHG emissions and produce more than 70% of the world's solid waste, which is biodegradable. The third-largest global source of methane gas emissions is the landfill. Pakistan's methane emissions rank 135th globally and account for approximately 0.8 percent of all greenhouse gas emissions. Human health is importantly impacted by solid waste contamination. Unchecked SW accelerates the growth and reproduction of dengue .

2.8 URBAN PLANNING

Urban planners' plans of replacing current social facilities like homes, schools, and hospitals with more people have failed. This has led to an increase in resident communities, whose geographic locations are surrounded by unfavorable conditions that make for bad living conditions for people. These studies make clear that a lack of information, a negative attitude on the part of the inhabitants, and poor governance are to blame for the hostile state of Kibera, the largest slum in Africa, which is located in Nairobi, Kenya. Kibera is a part of Nairobi, which has a population of 6 million. Between 500,000 and 700,000 people are thought to live in the 225-hectare community as a whole. Government statistics show that 82% of Kenyans reside in underdeveloped neighborhoods and rely on charcoal for energy (Karekezi, 2002). Because of their reliance on fuelwood for their livelihood, Kenya's forests are rapidly disappearing, which has a negative impact on the local climate, wild flora and animals, water supplies, and forest dwellers. The majority of people in Kibera live on less than $1 a day. Of them, half are unemloyed. An average of 205 metric tonnes of waste are produced each day and 75,000 tonnes annually in Kibera. An adequate waste collection system was lacking, and as a result, half of the waste was dumped is, which was dangerous to the people's human health. In Kibera, the prevalence of diarrhea is 40%, more than three times greater than the rest of the city, and is particularly prevalent in children under the age of three.

People in the developed world have been promoting the idea of diverting waste materials since the 1990s with the catchphrase "Reduce, Reuse, Recycle," commonly known as the 3Rs. The use of this expression integrates a practical approach to waste management because every technique has environmental advantages, ranging from the most important to the least important. The first two Rs are given priority from an environmental standpoint since they address waste-related problems through prevention. Reduce and reuse techniques help to save the environment while saving money, reducing land pollution, and using less energy. The first of the three approaches, "Reduce," is chosen because it is the most effective means of achieving the greatest reduction in solid waste. Waste minimization and waste prevention are other names for this. There are numerous techniques to reduce waste production. Source reduction is one strategy that aims to reduce waste by doing it in the manufacturing process itself. As a result, source reduction maximizes energy efficiency because it produces fewer output materials and cuts out laborious disposal procedures. Econoenefits will also come from using less fewer resources. This strategy works well for both individuals and businesses. For instance, early computers were big and took up a lot of space, taking up several square feet. Due to the usage of composite materials, current computers are now light, compact, and manageable. Another illustration is when a teacher provides her students with two pages of handout material; by giving them double-sided photocopies, she cuts the amount of paper used by 50%. Over time, there will likely be a decrease in the overall energy consumed by the machine to photocopy them. The use of source reduction in the production process will lead to a decrease in the output materials used for packaging. New packaging materials may therefore offer the same protection to the product while using the least amount of resources. Another strategy is to produce things that are simple to compost, reform, and repair. We need to impose a fee on each consumer for the amount of solid waste they deposit on public property. We need to offer a free waste collection service for reusable and recyclable goods. We need to follow the lead of several industrialized countries and enact cradle-to-grave accountability laws that oblige manufacturing facilities to recover a variety of old goods, including automobiles and other electronic gadgets.

2.9 WASTE REDUCTION

Waste reduction decreases pollution, saves money, reduces material consumption, and boosts the economy by providing jobs. More material can be produced with the same amount of resources and energy. For instance, the old information can be written over a used letter envelope to reuse it. Reusing materials might occasionally require expensive repairs that require time, work, and energy. For instance, reusing a disposable polystyrene cup more than once requires washing the cup, which produces some wastewater, and adds to the energy cost. Reuse is an effective and widespread techque in many developing countries.

Figure 2.7. Wat Pa Maha Chedi Kaew (Thai: วัดป่ามหาเจดีย์แก้ว; RTGS: Wat Pa Maha Chedi Kaeo, literally *Wilde the Great Glass Pagoda*)Buddhist temple. (Source: By Mark Fischer - Million Bottle Temple, CC BY-SA 2.0, https://commons.wa/commons/7/7f/Grand_Palace_Bangkok.jpg, Creative Commons License).

The government constructed Buddhist temples in Northeast Thailand using a lot of beer ottles (Figure 2.7). The brightly coured used bottles enhance the temples' aesthetics and let light into the interiors. Bottle caps were recycled to create mosaic artwork. Recycling is the process of sorting waste materials into input materials that are then used to make new products. In order to survive, it is a widespread practice in underdeveloped countries to recycle waste informally. Waste pickers are members of vulnerable groups in our society, including the elderly, the disabled, women, children, and new immigrants. They typically work in open landfills and on the streets, where they frequently come into contact with various types of solid waste that pose serious health concerns, and they live in a dangerous and unsanitary environment.

Lahore recycles around 27% of its waste in an improper manner(Azam et al., 2020). There aren't any active waste disposal facilities that use formal recycling techniques at the moment. This city is not an example of how well the government performs in the area of waste management. However, waste collectors gather wasted paper for recycling in the pulp business. Recycling in Indonesia reduces waste generation by about 10%. Scavengers are also essential to Iran's efforts to reduce SW. Enhancing recycling and lowering waste production have been made possible by numerous initiatives likevironmental education and training. A pPromising evidence of public support for recycling is the commercial sector's growing readiness to do so. The German nonprofit Deutsche Gesellschaft für Internationale Zusammenarbeit is collaborating with recycling company Taka Taka Solutions in Nairobi, Kenya, to recycle an amazing 95 percent of the 40 tonnes of waste produced daily. Contrarily, customers are more likely to purchase things that are simple to recycle. Small household equipment that assists recycling and minimizes waste volume have been bought by some consumers. Living patterns, traditions and national cultures, and consumption of commodities impact the generation and composition of domestic waste.

Given the rising global volume of solid waste, it is obvious that waste-related problscend national boundaries. All emergingdeveloping countries must work together internationally to find solutions to these problems. Although some international accords are regional, they must be global. Developing countries must follow their footprint in the direction of sustainable development, just as rich countries like the European Union and the United States have a number of policies about the reduction of solid wastes(Scarlat, Fahl, & Dallemand, 2019). Furthermore, since it poses ethical problems, steps should be taken to prevent the transfer of dangerous solid minerals from richer to poorer countries. International collaboration to fight the solid waste problems caused by a lack of environmental rules is becoming more and more popular. The developing world's population increase and economic progress have given rise to new problems in the waste management industry. Nanotechnology adds to the complexity of products, necessitating more sophisticated waste management procedures. There is a lack of sufficiently accurate and reliable data on waste generation. Therefore, there is a need for the stakeholders involved in waste management or making decisions about it to have a far better grasp of waste characteristics. Planning for a smaller ecological and carbon footprint as well as changing the ways in which resources are consumed are necessary for achieving sustainable development in developing ces. By 2050, it's predicted that lower lower-middle-income countries will increase SWG to 0.79 kg per capita per day, surpassing low-income countries. The growing material footprint of developing countries shows profound shifts in lifestyle expectations. In these countries, solid waste management im is consistently poor. The community's overall waste generating problems can be successfully projected by international and national nongovernmental organizations.

We must take the proper community action, grant rights to individuals working in the waste management industry, and devise workable solutions to keep people in their current jobs. International treaties and EPA regulations are required to safeguard the most vulnerable countries from exploitation. SWG is a new and pressing conceveloping countries. To achieve emergingdeveloping countries free of waste, more research is needed to assess the true potential of hazardous waste release from industry and to maximize recycling. The tools needed to accomplish the aims include environmental education and human resource development. Different waste management methods are used in industrialized and developing countries. Any chemical reaction or process will also produce waste, which is an undesired or unusable substance. Every industry produces waste, which is categorized based on where it came from. Solid or municipal solid waste is the general term used to describe the overall amount of waste. Each kind of waste has an impact on the soil, air, and water ecosystems. Municipal, biomedical, hazardous, and particularly hazardous waste are the types of waste. Agriculture, steel, her, cement, textile, sugar, plastic, tyretire, battery, and automobile manufacturing are the major industries that contribute to waste generation. Recycling, bioconversion, and incineration are the three main methods for managing waste in poor countries. However, these methods are sometimes used improperly for a variety of reasons. Open burning of waste results in many harmful aerosols of SO, CO, NO, CO_2, PM and other pollutant emissions, and open disposal of waste at the end has major negative effects on the environment and human health (Cheng et al., 2020). Solid waste also contaminates the ecosystem and water sources. The primary concern of social and environmental repercussions is the improper management of solid waste or waste. The improper and unplanned disposal of waste contributes importantly to the pollution of the soil, water, and ecosystem. Surface water pollutionso a result of improper dumping and their unchecked flow. Another important flaw in poor countries is the absence of effective waste management methods. The management of waste presents a number of difficulties for developing countries, including a lack of appropriate structure and prior regulations for various waste fragments/ components. In developing countries, there is a large amount of waste that is not adequately segregated and is either burned or thrown in the open, which causes various sorts of pollution or the spread of those pollutions.

In underdeveloped countries, a number of variables that aren't functioning effectively lead to poor waste management. These include infrastructure, finance, education, and the resources they require.

2.10 MANAGEMENT

Waste management is a problem for the environment, primarily because of open burning and dumping. Although all waste management solutions in developing countries are essentiallse in affluent countries, these countries' the approaches to waste management in these countries fall shorof expectations. Waste management programmes are challenging everywhere in the world, but in less developed countries these problems are even more severe. The main obstacl to a successful waste management programme are a lack of preventive measures, inadequate waste management technologies and equipment, a a lack of investment money.

This programme is deteriorating primarily owing to citizen ignorance, which ultimately causes the environment to deteriorate due to waste dumping, burning, and poor management, as well as economic interdependenThe nature, consistency, and sources of the hazardous waste only make the situation worse. Many multinational corporations relocate their facilities to less developed countries, and those countries have outlawed the usage of technologies. Transporters, disposal sites, and scavengers are the next largest sources of waste production after industry. Depending on the amount and type of waste produced, waste management in villages and cities varies greatly in developing countries. It is important to take note of these variations. Heavy metal pollution of the soil, water, and vegetation is caused by the improper disposal of waste. Lack of the fundamental functional components of waste management, storage, collection, transportation, and treatment leads to poor waste management in all industries. Open burning produces CO, CO_2, SO, NO, and PM_{10}, which has an impact on the environment and poses health concerns to those who work there. Waste management, which needs to be improved, is the outcome of numerous social and environmental effects. Waste in an industry refers to any materials that are left over or underused. This is the stuff that gets thrown away after it's been put to use. By slightly or importantly raising their economic value, we can turn waste into a byproduct. Examples include water waste and municipal solid waste. The developing countries are those with poor or intermediate incomes, and the amount of waste produced worldwide is relatively large. The pace of waste production is anticipated to increase by around 20% per year from 2016 levels to 3.40 billion tonnes in 2050 (Chatterjee & Sui, 2019). More than 90% of waste in poor countries is either burnt off or discarded. These actions have negative effects on the environment, safety, and health. Through the production of pollutants, improperly handled waste contributes to global climate change, as well as increasing or producing diseases. A better environment requires proper waste management, which is a major concern for developing countries. For instance, adequate waste management is prohibitively expensive, requiring 20 to 50 percent of the factory's budget. Waste mismanagement is made worse by a wide range of social, cultural, economic, and governmental problems. As it involves important budgetary expenditure, MSW management is the first problem that must be addressed before other services can be offered to the public. Mishandled waste has a number of harmful repercussions on the economy, human health, and global climate. Waste management becomes more expensive when it is not done properly. Collectively, South Asia, East Asia, and the Pacific region produce close to 1 MT of waste every day. When compared to high-income countries, low-income countries spend more money on waste collection than disposal.

Figure 2.8. Paper waste. (Source: https://upload.wikimedia.org/wikipedia/commons/7/7a/20080826_Waste_Paper.jpg, Creative Commons License).

Paper and organic waste are both present in Asia, where waste is growing most quickly (Figure 2.8). The global generation of waste is rising. For instance, the cities produced 2.01 billion tonnes of solid waste in 2016, or 0.74 kg per person per day. However, due to the increased population in 2050, this annual waste production will climb to 3.40 billion tonnes. The primary method of waste management in underdeveloped countries is the burning or open disposal of generated waste, which has important negative effects on the environment and huan health. Every time waste treatment technology is used, it degrades the soil, water, and air quality.

While poor countries produce mixed waste, the majority of affluent countries produce distinct hazardous waste. Hazardous solid waste came primarily from the informal sector, which is the main problem in developing countries with small, labor-intensive businesses. In contrast to the industrialized world, the misuse of dangerous agrochemicals leads to more severe situations in developing countries. Crop and animal wastes are produced during crop cultivation, animal husbandry, and agricultural processing; these residues are frequently recycled or reused as soil amendments and animal feed. Due to improper treatment, several hazardous waste components, including fertilizers, pesticides, veterinary products, and animal remains, pose a serious threat to human health and the environment. Most pesticides and fertilizers used in poor countries end up being added to the surface and groundwater, havs (Tilman et al, Cassman, Matson, Naylor, & Polasky, 2002). The rapid rise in municipal solid waste in low-income countries is mostly caused by population growth and economic expansion. At every stage of waste management, this generated MSW is iroperly managed and creates important problems.

Open dumping, (Figure 2.9) incorrect waste storage, ignorance, and improper placement of pickup bins are the main problems with MSW. Household, business, and demolition waste made up the municipal waste. In Pakistan, household waste production averages 0.79 kilograms per capita per day.

Figure 2.9. Open dumping. (Source: Mohamed NA, Asfaha YG, Wachemo AC. Integration of Multicriteria Decision Analysis and GIS for Evaluating the Site Suitability for the Landfill in Hargeisa City and Its Environs, Somaliland. *Sustainability*. 2in_santiketan%2C_khoai%2C_Birbhum%2C_West_Bengal.jpg, Creative Commons License).

Disease and pollution concerns are increased in low-income countries as a result of improper management of household waste. The best way to manage household waste is either via safe disposal after combustion or another method. The technique's strengths and shortcomings determine how effective it is.

There is no one strategy that works for all types of waste. The ideal way to handle household waste is to properly teac the staff and separate the waste into distinct coloured bins before disposing of it in the right spot. Household waste makes up a portion of municipal solid waste that is improperly managed in underdeveloped countries. Household waste is the category for waste generated in healthcare institutions. About 75 to 90 percent of domestic waste is non-hazardous, and the remaining 10 to 25 percent poses health hazards. The majority of household waste is discarded in underdeveloped countries. Although some places adopt improved safety management. Open dumping is a common example of poor waste management in households. Waste pickers approach the waste at open dumping sites to collect the recyclable materials, but doing so spreads infectious diseases. Another waste management error/problem is uncontrolled burning, which releases pollutants and poisons similar to PCDD. Further, 82.2 percent of the waste in Palestine originated from homes, and 17.9 percent came from hospita. Because of the infiltration process, waste disposals' leachates affect groundwater. Another place where domestic waste is improperly managed is Bangladesh, where after collection, everything is either burned or dumped into the public sewer system. Many infectious diseases are caused by this technique. Water used for cleaning, agricultural, and domestic activities during the rainy season is contaminated with leachate from disposal sites.

2.11 URBAN WASTE SITUATION

Urban areas are under more population pressure since they have more facilities than rural places. Construction and demolition waste generated during the building of new dwellings and shelters must be properly managed. Iron, concrete, plastic, and ceramics are produced as waste during construction and building destruction. The majority of them can be recycled or utilized again, particularly for waste that contains plastic and iron. the waste that scavengers collected from industries. The majority of the waste is discarded in the open on the streend the remainder is put in inappropriate locations. The sSerious health risks for people are the result of this improper waste management. Burning plastic releases harmful gases and other pollutants into the air. Mismanaged sanitation during the construction of buildings or dwellings leads to sewage waste, which ultimately pollutes freshwater, subterranean water, and sanitary landfills. The handling of hazardous waste is hampered by public ignorance. Developing countries suffer from improper waste management, which worsens as the types, sources, and quantities of waste expand. In underdeveloped countries, where the risk is higher than in developed countries, all industries produce hazardous waste. Multinational corporations relocated their manufacturing facilities to less developed countries where they exploited technologies that were forbidden in their home countries. Every industry produces industrial waste, and much of it is improperly managed and disposed of. Surface runoff and leachate from these dumping grounds further contaminate the water.

Agriculture accounts for the majority of the GDP in developing countries. Farmers utilize fertilizers, pesticides, insecticides, fungicides, weedicides, and other chemicals in the agriculture industry. The chemicals and equipment they require increase the waste from these sources. The main waste sources include crop bran, straw, crop remnants, animal carcasses, and animal feces. These wastes must be effectively managed. Crop residues are improperly managed since they are mostly burned for energy rather than composted or turned into green manure, which results in the production of deadly gases and air pollution. When animal carcasses are left onround to decompose, various bacterial and fungal growth resultoccurs, which has an impact on human health. The accumulation of animal waste releases an unpleasant odor and depletes the organic materials exposed to the sun (Saravanan et al., 2021). The quality of the land, water, air, and human health is impacted by improper fertilizer and pesticide application. There are many waste products that are recycled in the cement business, but the primary waste material that can degrade air quality is cement dust. It speeds up the breakdown process and eliminates parasites and bacteria. Another important contaminant or waste produced by the cement industry is CO_2. The two main contributors to CO_2 production are combustion and fuel consumion. The total CO_2 produced by the cement industry is 5%.

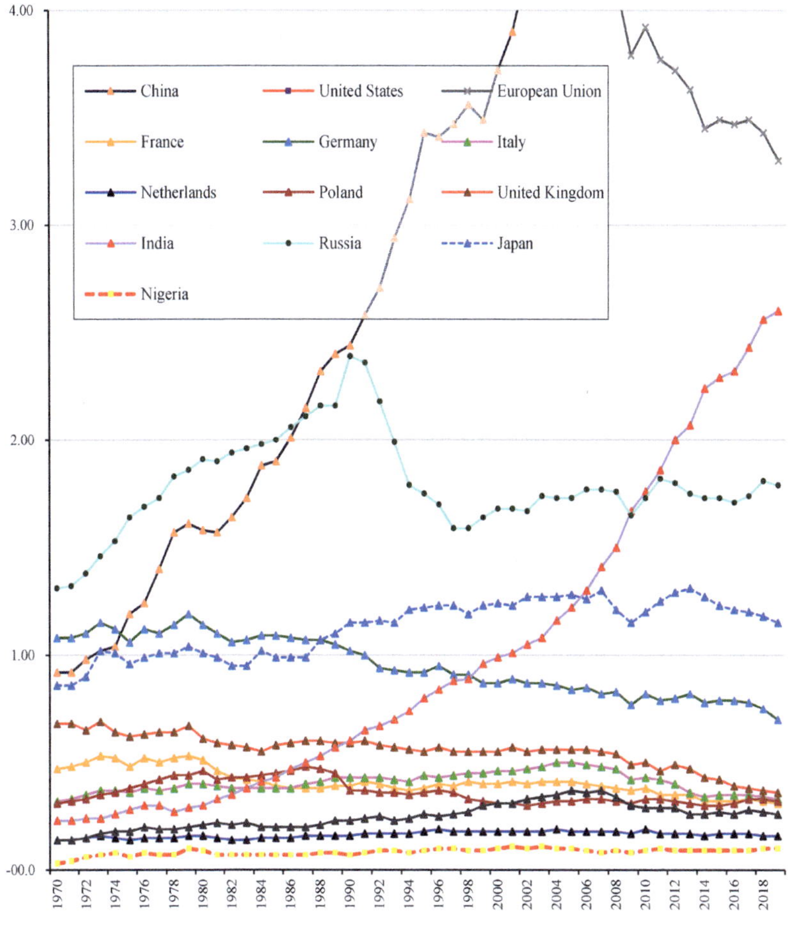

Figure 2.10. Carbon-dioxide (CO70 to 2019 in Nigeria and other countries. Modern cement plants. (Source: Etim M-A, Babaremu K, Lazarus J, Omole D. Health Risk and Environmental Assessment of Cement Production in Nigeria. *Atmos*.wikimedia.org/wikipedia/commons/5/5c/Cement-factoCreative Commons License).

Modern cement plants (Figure 2.10) generate 60% of their CO_2 from the thermal treatment of $CaCO_3$, 30% from the combustion of fuel in the furnace, and 10% from other plant processes (Figure 2.10). Depending on the type of plastic and the chemicals used to make it, waste in the plastics sector is often recycled or repurposed. Minimization, recycling, and the use of biodegradable plastics are the best methods for managing plastic waste. However, burning, dumping, and open disposal of plastic waste are examples of poor management of the material in underdeveloped countries. Burning plastic waste improperly results in smoke that has harmful consequences on people, animals, and the environment. Chlorinated plastic deteriorates soil, contaminating water through seepage and ultimately disrupting the activity of bacteria and other microbes. Methane, a potent greenhouse gas, is occasionally released by biodegradable plastic, contributing to global warming. Due to its carcinogenic effects, diethylhexyl phthalate contaminates the ocean, which ultimately contaminates the food chain. Plastic is used as building material. Plastic recycling varies by country and polymer type. Polyethylene terephthalate and high-density polyethylene are recycled more often than other types of plastic. Aroun0% of developing countries recycle their plastic waste.

Waste tyres tires are not effectively managed in underdeveloped ces due to ineffective management systems. The majority of used tyres tires are disposed of or used as fuel in s businesses, such as the cement industry. The recovered waste tyres tires are mostly used by scavengers, scrap merchants and recovery businesses. Due to their robust composition and design and the potentially toxic materials they contain, their improper management poses a serious risk to the environment. Shere are no facilities for the processingreatment of used tyres tires in underdeveloped countries, these tyres tires are either burned or piled up at dissites. There are many methods for effectively controlling waste tyres tires that, when correctly implemented, canth efficient and secure. The most efficient ways to manage used tyres tires are incineration for energy production, landfilling, pyrolysis, scale reuse, grinding for various purposes, and retreading. Waste tyres tires are of large volume, thus impranagement reduces the number of sanitary landfills. Burning used tyres tires damages the environmed raises health hazards for the general public. In several dumps, tyre tire waste multiplies diseasors. Another example of a disease that can spread because of used tyres tires is dengue. The SO_2, CO, poly-aromatic hydrocarbons, NO_2 volatile organcreased by burning tyrestires (Curtis, Rea, Smith-Willis, Fenyves, & Pan et al, 2006). Open burning and dumping are the two major inefficient methods of disposing of spent batteries. Leachates with high concentrations of heavy metals caused by battery disposal degrade the land and water ecosystem. Burning used batteries similarly results in the production of PCDD/F, BC, and PM. All living things, including humans, are affected by the heavy metals Hg, Pb, Cd, and Mn. In poor countries, burning is a common and poorly supervised process for recovering various metals. For instance, following import without separation, 10,000 tonnes of batteries were dumped in landfills in Iran. The disposal of used batteries in landfills results in an increase in the concentration of hazardous metals like Cd, Li, Ni, and Ar. Because there is inadequate infrastructure and no specific rules and regulations for this type of waste maaste batteries are improperly managed.

2.12 AUTOMOBILE SECTOR

Automobile The automobile sector of the economy produces a variety of waste, including solvents, metals, batteries, plastic, and glass. Coke, coal dust, and refractory waste are produced in the iron/steel sector. For every tonne of steel produced in India, 1.2 tonnes of waste are produced (Yellishetty, Ranjith, & Tharumarajah, 2010). 63 percent of the waste from the steel industry that needs to be recycled or utilized gets dumped. Ferruginous and non-ferruginous waste are produced in the steel industry.

The ferruginous waste, which contains iron, comes from steel melting facilities, blast kilns, flue dust, sludge, etc. The sugar industry produces solid waste such as bagasse, and leaf debris, but it also produces liquid waste from washing through processing that must be treated before being added to water reservoirs to prevent water pollution. Sugarcane bagasse has a high lignin, cellulose, and hemicellulose content. The manufacturing of textiles, pulp and paper, animal feed, fuel, food, chemicals, enzymes, and polymers are among the industries that use bagasse most effectively. Bagasse, a solid waste, was used in China as fuel for boilers. The waste product known as sugarcane filter cake is produced when the sugar cane juice is diluted and has an adequate level of nitrogen. Sugar industry waste is a major environmental problem in countries like China, Brazil, and India. The effluent from cane molasses distilleries is also found in wastewater. One tonne of sugar is thought to require 20–30 tonnes of water for processing. Bagasse, OM, suspended solids, sludge, effluent, and filter cake are among the waste products from the sugar cane industry. India creates 1000 L of effluent from processing 1 tonne of sugar cane. When sugarcane effluent is improperly managed, such as by using it directly for irrigation, growth, yield, and soil health are all negatively impacted. The presence of hazardous chemicals including Pb, Zn, Fe, Mn, and Cu causes the effluent to harm the paddy and fish. Sugar cane leaves are improperly handled because open burning creates fly ash. The microbiological activity in the soil is badly damaged by fly ash, and it also pollutes the air, causing ailments of the lungs.

Due to chemical reactions, waste is produced at every stage of the leather production process. Water effluent, which primarily contains lead, chromium, cadmium, nickel, and zinc, is improperly managed and dumped into water bodies without being treated. It is commonly known that only 75 kg of leather can be made from 500 kg of skins, with the remainder considered waste. When waste is disposed of in landfills, dangerous bacteria may be produced. The water discharge from the leather industry is to blame for one-third of all water contamination. The primary ingredient in the textile business, known as effluent, is wastewater. It results from the washing of fiber to the fished product. 200L of water is needed for the manufacturing of 1 kilogramme of textiles. The environment is harmed by untreated effluent. The wet processing results in the largest amount of wastewater production. While detergents, surfactants, emulsifiers, and dispersants are the main causes of aquatic toxicity, dyes generate low aquatic toxicity. The procedure of wet finishing treatment in the textile bath results in mixed wastewater. The primary source of pollution is the discharge of textile baths into the environment. The byproducts of textile processing, such as grain dust, sawdust, fiber dust, and coal dust, harm workers when inhaled. After combining with the air, this dust contaminates the atmosphere with suspended solids. As a result of breathing air that has been contaminated with cotton, respiratory ailments increase. Small-scale textile enterprisesin poor countries still lack the infrastructure to prevent such mismanagements.

2.13 BIOLOGICAL WASTE

The biological waste is generated by various healthcare facilities and poses a serious health concern to people. Different chemicals, biohazardous wastes, pharmaceutical wastes, radioactive substances, pathological wastes, and genotoxic wastes are some of the categories of biomedical waste that have an impact on both human health and the environment. Hospital wastes have a 10–25% negative impact on people, including patients, healthcare professionals, and other people (Padmanabhan & Barik, 2019). Hospital wastes are frequently dangerous and contagious, and their improper treatment feeds the spread of many diseases and degrades the environment. Hospital waste is a major cause of the spread of dangerous diseases like cholera, TB, typhoid, hepatitis, HIV, and AIDS. Globally, hospitals produce 0.5 to 2.0 kilograms of waste per patient every day. Hospital waste management problems in Pakistan are posing major risks. Hospital waste in Pakistan ranges from 0.1 to 0.7 kilograms. Hospital waste becomes entirely contagious when the infected and non-infectious portions of waste are combined together due to improper disposal. The risk of infections among waste collectors is increased by the improper disposal of this waste. The spread of disease in Pakistan is a result of poor waste management practices, a lack of resources, and inadequate infrastructure. Hospital waste is adequately managed using the incinerator method, which is typically absent in underdeveloped countries and leads to open dumping in various parts of the country. Nursing homes, clinics, pharmacies, medical facilities, and outpatient surgery centers all produce waste that ends up in hospitals. By contaminating the water supply, improper biomedical waste management poses a serious threat to society and the environment. The improper handling of waste at every stage of waste management, including generaion, collection, storage, transport, and treatment, constitutes clinical waste mal-administration. Improper storage, improper handling of biological waste, and improper covering of the waste during transportation are all examples of improper handling. These techniques also draw flies and rodents. Due to environmental concerns, the practice of incineration is likewise transient. Dioxins, greenhouse gases, and other harmful emissions are produced when medical waste is burned indoors or outdoors, and these emissions can increase global temperatures, lead to cancer, and worsen respiratory conditions.

Explosive, radioactive, and electric or electronic waste are the three types of special hazardous waste (Dehghani, Omrani, & Karri, 2021). There are several handling technologies for each category of special hazardous waste. Propellers, weapon ballistics, bulk explosives, consumer and commercial explosives, munitions, roadside and signal flares, marine, missile propellants, and vehicle airbag propellants are some of the waste products in the explosive category. Fluid movement is produced with a propellant. The majority of propellants are burned; the rest degrade. Munition, which includes propellants, explosives, and fireworks in various forms like bombs, mines, ammo, and grenades, is a type of explosive waste. Damaged, underused, or dysfunctional fireworks produce more hazardous waste as a result of their activities. High energy can be produced with very little waste and a very tiny amount of fuel. However, because this waste is so radioactive, it must be handled with caution. Nuclear waste must be appropriately disposed of during the electricity producing process. Agriculture, health, research, mineral exploration, non-destructive testing, and manufacturing have all been shown to benefit from the use of radioactive waste.

Waste that cannot be recycled in industrialized countries is either disposed of in landfills, burned, or shipped to underdeveloped countries. 80 percent of the e-waste sent from the US went to Asia, and then 90 percent went to China. In addition to cadmium, mercury, antimony, and beryllium, computer waste also contains polymers, iron, aluminum, copper, lead, zinc, and tin. Early PC generations contained 4 g of gold; however modern PCs only contain 1g. Lead now makes up a larger portion of electronic waste than other metals. Inadequate control, improper regulation, and various effects on the environment and human health are the main causes of the poor waste management practices that pose serious pollution concerns to the general public. The federal and national agencies are encouraged to implement new infrastructure for the management of hazardous and unstable waste because of inadequate management. The various waste management techniques, such as avoidance, waste reuse, recycling, and compositing, are encouraged. The decision on which waste management techniques to use is based on the technical and economic factors of the waste that needs to be treated. Open burning, open dumping, and recycling are the methods of wate management most commonly used in underdeveloped countries. This includes all types of waste and unprofessional handling of waste by waste pickers, which transmits severe intimidation to the ennment and human life. All of these practices involve unsustainable and unprofessional ways for of handling waste, which have worsened the environmental problems. In any case, while some advanced practices are present in developing countries, those technologies are still unavailable in the least developed countries. The strongest predictor of waste management tactics is a country's economic growth, but the most important component for effective waste management is a country's population behavior. Open dumping is a common waste management method in developing countries, where unregulated locations and uncontrolled disposal make the problem worse. Especially in densely populated cities, this poorly managed waste disposal would result in serious health problems, surface and subsurface water pollution, and air contamination.

Another typical method of waste management in poor countries is waste burning. It is a primary contributor to air pollution, which poses risks to the environment and raisehazards. According to the U.S. Environmental Protection Agency (USEPA), burning waste resulted results in the release of GHG into the atmosphere (Broun & Sattler, 2016)reduction of methane emissions from municipal solid waste (MSW. Burning diverse wastes is an important source of toxins being released into the environment, such as particulate matter and volatile organic compounds when burning plastic waste. Around the world, between 40 and 50 percent of municipal solid waste is burned in the open unrestrictedly, and the countries with the highest recorded emissions from burning waste are China, India, Brazil, Mexico, Pakistan, and Turkey. The addion of the informal sector in waste collection could support the formal waste management programme in low-income countries where the formal collecting system of waste has not yet been built. The logical process that informal pickers adhere to is as follows: first, gather waste from the location, then sell it to a merchant or to the trading locations. The trade points additionally sell it to formal or informal recyclers, as well as directly to the producers. The chain is finished in this manner, as shown in several case studies. The operations of the informal sector are consistent with the concepts of the circular economy since they tend to reduce the amount of waste being discharged and immediately recover the material. According to case studies from numerous developing countries, including India, Bosnia, Malawi, Colombia, and many others, the informal sector is crucial to waste management. To restore environmental sustainability, the informal sector may have joined the formal stor as an active participant. According to numerous studies, waste rerouting or transfer programmes, compost and biogas production, and execution of waste to energy plans and skill tools can all improve solid waste management in developing countries. Recycling of metals, glass, and other materials can help reduce the amount of waste that is converted into energy. Environmental contamination is still a grave concern on a global scale.

2.14 URBAN EXPANSION

Megacities are the result of extensive urban development and expansion that takes place globally (Shafizadeh-Moghadam & Helbich, 2015)and particularly in Mumbai, India, where growth is very pronounced. To cope with the planning challenges this will bring, developing a retrospective understanding of urban land-use dynamics and the underlying driving-forces behind urban growth is a key prerequisite. This research uses regression-based land-use change models – and in particular non-spatial logistic regression models. Certain driving forces encourage poor management of urban dimensions. These pressures include the growing urbanization and industrialization, incorrect public behavior, and a government that is indifferent, especially in terms of money allocation. All of these motivating factors put pressure on urban growth, resulting in social fragmentation and creating unnatural social dynamics. Rapid urbanization also leads to a lack of the necessary foundational infrastructure and improper management of such urban features. Because poor waste management in megacities is closely related to inadequate management of major metropolitan growth. Therefore, it is crucial to research factors proportionally accountable for waste mishandling and urban corrupt governance. There is a global problem with waste management. On a larger scale, the effects of this poor management include environmental damage, social exclusion, and a lack of economic stability. The resolution of this worllem necessitates comprehensive, coordinated analyses and methods. However, it is crucial to recognise recognize the divergences between the variables causing waste mismanagement in rich and underdeveloped countries. While some causal elements are universal to developed countries and need to be addressed in a similar but sustainable manner, others are exclusive to developing countries. Social, financial, economic, technological, and management or government-related variables are just a few that could make managing waste in urban and rural areas more difficult.

In addition, because of outdated infrastructure and a lack of competent and consistent maintenance, current infrastructure frequently fails to ensure safe disposal. It causes the emergence of unsafe circumstances, which further increases societal vulnerability. Health problems and other related problems eventually become a result of these infrastructure deficiencies, ongoing exposure to unsafe conditions, diminished capacity, and potentiality. People experience important time, financial, and labor costs as a result of these obstacles when managing their own wastes and while preventing and treating infectious diseases. The effects of environmental risks brought on by improper sewage and wastewater management are much more damaging and difficult to evaluate. One of the difficulties in low-income countries has been the economic crisis and the improper and insufficient release of government funding designated for waste management. Metropolitan waste collection and disposal have not been successful due to old-fashioned waste management equipment that has not been maintained and ack of funding for their timely nt. The COVID-19 pandemic has caused an unexpected collapse of in waste management worldwide. Pandemic The pandemic and economic collapse have had an adverse effect on the world. Our waste management system urgently needs to incorporate assessment, readiness, and resilience capabilities so that it can handle residential and medical waste and contain infectious diseases in the event of such a calamity. Public perception and awareness importantly contributed to waste mismanagement in low-income countries. The general populace is unaware of the importance of properly disposing of waste (Abd'Razack, Medayese, Shaibu, & Adeleye, 2017)Medayese, Shaibu, & Adeleye, 2017. They blatantly discard the waste in open drains, canals with running water, or other unapproved locations that pose threats to human health. It demonstrates how sanitation law enforcement organizations failed to implement the law. Planning for waste management is a subset of planning for the environment.

In addition to other causal variables, infrastructure stress, outdated technological equipmen of cutting-edge research and technology all contribute to the improper management of waste. Government's The incapacity of governments to bring high-end research and its inability to distribute sufficient cash are two of the contributing causes. Since the dawn of humanity, waste has been connected to humans. As the population expanded, dealing with waste on our planet became a major task. With advancements in science and technology, the amount of waste that is introduced into the environment has greatly increased. On the other side, science is finding it difficult to meet this challenge. The countries with high revenues have managed to protect their populations and demographic environments from the harmful effects of waste in some way. Because of poor waste disposal plans, infrastructure, and a troubled economy, the problem of waste mismanagemnt is especially severe in less developed and developing countries. Regardless of geographic limits, this poor and unsustainable waste management has quickly led to pollution of the soil, water, and atmosphere with the worst effects on human health. Europe has faced ongoing challenges from several diseases that in the past were caused by water contamination. The effects of environmental contamination, particularly on water-borne diseases, are more pronounced in the twenty-first century, with important effects on Africa and Asia. Despite new technological advancements, improved waste management laws, enlightened human health oversight, and deteriorating waste treatment and disposal, individuals have a relatively low level of technological adoption.

Due to high environmental pollution brought on by improper waste management and unscientific waste disposal, the air, water, and soil become more polluted. Waste and wastewater severely damage the local flora and fauna as well as the health of the population. The biosphere is already at risk due to the unchecked discharge of pollutants into the air and soil. The threat to food safety and security is posed by the contamination of land and water resources. Soil is essential for life since it contains both minerals and organic materials. Soil is lifeless without its biota. Changes in soil composition and the eradication of soil life seriously harm the biosphere as a whole. Soil contamination is mostly caused by improper agricultural practices, sewage water managemd industrial waste disposal. Up until 2025, the global production of waste from metropolitan areas should is predicted to range from 1300 to 2200 million tonnes annually (Pujara, Pathak, Sharma, & Govani, 2019)apart from the metro-cities. This practice poses significant environmental and health risks due to toxic and greenhouse gases (GHGs. In underdeveloped countries, where 219 kg of waste were produced per capita per year in 2010, this waste output is modest. Because of this, this ecosystem receives an enormous volume of waste each year. In a similar manner, the production of waste will increase to 343 kg/person/year in 2025. The composition and characteristics of waste, which are in turn influenced by financial situation, technologies being adopted, living principles, food lifestyles, customs, learning rate, form of energy foundation, and climatic and geographical situations, are directly related to the environmental effects of waste. Due to poor waste management, concentrated heavy metals pose a serious risk because of their toxicity. Due to open dumping and discharge into the sewerage system, heavy metals find their way into both liquid and solid industrial effluents. In developing countries, industrial centers have previously come under scrutiny for their poor air quality. Megacities in underdeveloped countries are at the top of the list for air quality. Methane, CO_2, and several trace elements among the waste gases released by landfills have a considerable negative influence on both the environment anxtremely hypoxic decomposition, methane absorptions can reach up to 50% of the amount of gas. These smokes' This contribution to increased greenhouse gas influence and environmental change presents a challenge. Plastic waste poses the greatest threat to air quality of all the wastes. When burned on an open flame, plastic and rubber emit toxic gases into the air.

The main contaminants from poor waste management include persisanic contaminants, radionuclides, nutrients, heavy metals, lubricants, sediments, pathogens, etc. In emergingdeveloping countries, there is a greater prevalence of ocean waste dumping, particularly of plastic debris (Shahidul Islam & Tanaka, 2004)aquatic pollution did not receive much attention until a threshold level was reached with adverse consequences on the ecosystems and organisms. Aquatic pollution has become a global concern, but even so, most developing nations are still producing huge pollution loads and the trends are expected to increase. Knowledge of the pollution sources and impacts on ecosystems is important not only for a better understanding on the ecosystem responses to pollutants but also to formulate prevention measures. Many of the sources of aquatic pollutions are generally well known and huge effort has been devoted to the issue. Hwever, new concepts and ideas on environmental pollution are emerging (e.g., biological pollution. The marine fish and mammals have suffered damage. Plastic waste can be discovered in lakes, rivers, beaches, the center of the ocean, on the seafloor, and even inside aquatic species. There is a chance of creating a connection between plastics and the ways that invasive species pose a serious ecological hazard because of the abundant manifestation of microplastics, which has attracted the attention of ecotoxicologists. Because plastic takes generations to disintegrate, it does not break down quickly. Because of this, plastic tends to collect in water bodies, particularly seas, which is an important environmental problem for all coastal countries. There is an urgent need to reduce the flow of plastic and other waste into waters, despite the fact that this problem is now more well understood. The key determinants of the variation in the effects of waste on human health are the type of waste, its permanence, population exposure, and the availability of managemental solutions.

CHAPTER 3

Municipal Solid Waste Management

CONTENTS

3.1 INTRODUCTION

One of the environmental issues with the greatest global concern is solid waste. The environment has traditionally been seen as a drain for all the waste generated by human activities from the dawn of civilization. In the aftermath of urbanization and the rise of civilization, the technologically advanced society of today has significantly altered the natural environment, whether consciously or unconsciously. Human activities that serve everyday requirements contribute to all forms of pollution that affect the air, water, and land. These consequences are obvious since air and water are necessary for human existence. As a result, extensive efforts are being made to curb and avoid pollution. The effects of solid waste's contribution to environmental pollution are among the matters of greatest concern. Despite the reality that much more has to be done in developed countries to tackle the crisis, this is still being given much attention. It is a concern in underdeveloped nations. Waste management is frequently referred to as solid waste management. It is made up of materials that are left over after being used in daily life; examples include food waste and metals. Based on the source, origin, and kind of the waste, solid wastes are divided into a wide variety of categories. Municipal solid waste (Figure 3.1) is distinct from other sorts of waste in that its generation cannot be determined in time.

Figure 3.1. Municipal solid waste. (Source: https://upload.wikimedia.org/wikipedia/commons/e/e0/Municipal_solid_waste_in_Rome_03.jpg, Creative Commons License).

3.2 WASTE FROM DIFFERENT COMMUNITIES

Waste is produced at various locations at all times, and to manage the commotion and inconvenience of MSW, an organized process for waste storage, collection, segregation, transportation, processing, and disposal is being put into place through a controlled channel known as the MSW management system. Every country is progressing at a different pace to define and meet management targets while MSW is produced and treated globally (Vergara & Tchobanoglous, 2012). In developing nations, managing MSW presents significant difficulties due to the fast and rapid urbanization that is occurring. A vast amount of solid waste has been produced as a result of the increased range of products made available as a result of economic globalization. On the rate of waste generation, collection, coverage, storage, transport, or disposal volumes, there is no complete data. The management techniques used in developing countries are inadequate given the large waste generation. The majority of efforts to improve these systems have relied on importing Western technologies, however, these efforts have largely failed due to the inappropriateness of these imported technologies under the current situations. Theemergingdeveloping countries require a fundamental set of guidelines to manage their MSW. They create centers that provide a range of services, including employment opportunities, cafeterias, restaurants, and industries that produce a variety of goods. People from rural areas migrate to metropolitan areas in pursuit of a better standard of living and more money. Together with the metropolitan population, the floating population contributes significantly to waste production. The amount of waste produced depends on a number of variables, including the local population and consumption trends.

Each of these communities lives differently and eats differently. The amount and nature of a region's natural resources as well as peoples' eating and living patterns all influence waste production. These variables together generate a tremendous amount of waste, making the MSWM essential for metropolitan bodies. Since MSWM has special characteristics connected to constrained land areas and poor financial capability, it faces major managerial challenges in metropolitan areas of developing countries. Paper, plastic, metal, and glass collectively contribute an additional 30 percent or more by weight to the region's MSW generation rate, which is projected to be 0.45 kg per inhabitant per day, according to the Asian Development Bank Report on Solid Waste Management in the Pacific. According to the Human Development Report, it is not at all reassuring that developing and less developed countries have relatively easy access to water and sanitation.

45 percent of nations' populations lack access to adequate infrastructure and sanitation. The research also emphasizes that a growing portion of the population, more than 15% on average, is now residing on degraded property. As a result, governments will undoubtedly face serious challenges as a result of the effects of unplanned urbanization growth, particularly in terms of providing the need for adequate and reliable municipal services. As a result of the growth, waste generation will be of higher quality and complexity, placing a greater pressure on waste management systems that handle solid waste in general and MSW in particular. An interdisciplinary activity, MSW incorporates engineering concepts, economic and social concerns, as well as urban and regional planning as its foundation. From the time of generation till disposal, MSW is managed. The system needs resources in the form of labor and equipment for MSW handling operations, such as waste collection, transportation, loading, and unloading, in order to complete the extensive task of MSWM. For the final disposal of the waste, vehicles and large earthmoving equipment are required at landfill sites. The management of MSW should be planned, constructed, and carried out in a way that causes the least amount of environmental destruction possible.

3.3 MUNICIPAL SOLID WMANAGEMENT MSWM

Prudent application of heterogeneous technologies is required for effective and efficient Municipal Solid Waste Management (MSWM) in order to reduce environmental pollution (Pattnaik & Reddy, 2010). A committed group of professionals coordinates, plans, and implements the system while taking into account all factors. Mechanical engineering, for example, can be used to better handle and process equipment, chemical engineering, for better processes and technologies for effective treatment methods, and social science, for better public awareness and cooperation. In order to create sustainable economic models and to manage the system, economists and environmental engineers are also required. Above all, needs are satisfied by utilizing the tools and resources at hand. The management of MSW is dependent on the quantity and quality of the waste, and MSW characteristics differ from location to location. Food preferences, culture, climate, and other factors all contribute to this diversity. The amount of solid waste is influenced by gross domestic product per capita. Physical and chemical tests are used to evaluate the quality. Chemically, it describes the percentage composition of protein, carbohydrates, and lignin along with the elemental composition, that is, the percentage of carbon, nitrogen, and phosphorus. The industry needs adequate management because the population is growing daily and MSW production is inversely related to the population. However, MSW is seen as a low priority in this endeavor. The impact of MSW is indirect; rather, it arises from inappropriate handling of the waste, which endangers public health and diminishes aesthetic values.

A risk to public health is created by MSW dency 's ability to breed disease vectors, primarily flies and rats. Ineffective management causes insect and rodent infestations. The most obvious environmental damage is aesthetic, brought on by street waste and careless waste disposal. Despite being given less attention, solid waste is tied to all other forms of pollution. When waste is processed, like by incineration, poisonous vapors release into the air and pollute it. Leachate, produced by unmanaged and discarded waste at land eventually stainspollutes drinking water by washing out during the rainy season and floating down to nearby water bodies. Therefore, MSWM is a problem that also highlights a close connection to other types of pollution. MSWM is still relatively unknown in poor nations. Developing nations are unable to offer adequate solutions to the management challenge due to their restricted access to technology and resources. The risks of poe been recognised recognized in the majority of developing nations, and people have begun to consider improved technology and affordable, effective systems that would fit their needs. The ongoing connection between society and the system has assumed many different forms and is present in many different regions of the world. Integrated solid waste management is a comprehensive approach to waste prevention, recycling, and management. Thiextensive programme helps to safeguard both the environment and human health. In order to determine the best waste management strategy for a given situation, the Integrated Solid Waste Management (ISWM) evaluates the local requirement and conditions. Complete waste management, including composting, burning, and disposal at a well-constructed and managed landfill site, are the main activities. For national and local governments seeking to assure its effective and sustainable management, the enormous development in the quantity and variety of waste, due to ongoing economic growth, urbanization, and industrialization is becoming an increasing concern. The amount of MSW produced worldwide is continuously rising, and every government is now focusing on the best way to drastically minimize the management challenges associated with solid waste management. The global implementation of integrated MSW management is required. Various strategies and plans for carrying out MSWM are in use in industrialized nations. However, some industries are struggling to implement their integrated MSWM processes.

Major deficiencies in this area still need to be filled. According to the World Bank, developing nations devote 20% to 80% of their available resources on solid waste management (Altaf & Deshazo, 1996). Only 50% of the population benefits from waste collection and waste management systems since 30–60% of urban solid waste is not collected. Eighty to ninety percent of the overall budget for MSW management is spent on MSW collection in low-income nations. Countries with high incomes spend less than 10% of their overall budget. In developed nations, organized public participation lowers the cost of collection, promoting recycling and recovery.

The management of waste presents a significant problem for developing nations. The area that needs the most attention is how to reduce the total amount of waste while still making enough money. If the waste is directed toward material and resource recovery, the overall volume of waste might be decreased. Recovered resources and materials can thus be used to generate funding for waste management. This serves as the foundation for a 4R-based integrated municipal solid waste management system.

The justifications for implemeing the IMSWM programme in waste management procedures must be known and understood. Increases in a city's financial and developmental activities, which are fully dependent on lifestyle, affluence, the type of natural resources available, and consumption patterns, signal the expansion of a city. The features of waste have changed as consumer habits and lifestyles have changed substantially. MSW has become one of the most difficult issues of urban environmental deterioration as a result of our incapacity to properly comprehend the problem of waste generation.

A significant quantity of waste is handled by the informal sector without the necessary environmental safeguards, so it is absolutely necessary to incorporate them into the mainstream waste management process. An integrated strategy that includes collective management of all waste types and the adoption of the 4R policies and strategies is likely to be the best approach to waste management. The world population is expected to reach about 7.2 billion people in 2015, with everyday population growth. In fact, more than 150,000 additional people move ins every day in emergingdeveloping nations. Although uncontrolled and spontaneous growth are more likely to cause environmental issues than rapid urbanization, these issues include the encroachment of public space and riverbanks, air and water pollution, and the production of enormous volumes of solid waste.

3.4 RECYCLING

In urban areas, even a small increase in income can affect people's consumption habits, resulting in new waste forms and volumes that are harder for communities to manage. In the name of recycling research, developing and developed countries exhibit specific opposing characteristics. Because of their sophistited recycling programmes, developed countries have been discovered to conduct a number of industrialized recycling activities that are more or less detached from the daily lives of their citizens. Additionally, they concentrate on precise models, techniques, and policy analyses such as command-and-control, social psychology, and financial incentives. Less emphasis is placed on understanding recycling science as it is used in industrialized nations, and more emphasis is placed on practical, direct issues that have an impact on institutions and on MSWM fundamentals. Recycling systems are frequently used in developed nations to collect and sort waste for recycling. This type of work is low-tech, low-paying, and uncontrolled. The garbage that the rag pickers have collected is subsequently sold to exporters, middlemen, or recycling businesses.

Any item that has been fully utilized for one specific function and is no longer usable is referred to as waste. Waste is generated either by human beings, animals, plants, or by any natural or manmade activity. Waste can come in different shapes and sizes, including municipal solid waste, biodegradable waste, non-biodegradable waste, chemical waste, waste from building and demolition projects, electronic waste, biomedical waste, wastewater, sludge, hazardous waste, industrial waste, food waste, and more. Solid waste is any waste or debris that is discovered in a solid or semisolid physical state. Solid waste comes from a variety of places, including homes, public spaces, businesses, hospitals, industries, semisolid waste from wastewater treatment facilities, electronic companies, and more. MSW is waste made up of commonly used things for public use that have been discarded, including plastics, paper, green waste, electronic waste, and inert waste like building and demolition debris. Waste is first gathered and kept at a major storage point, such as street containers, from a variety of sources, including homes, businesses, street cleaning, and parks. After that, the waste is sent to secondary storage facilities in urban areas where it is held and separated before being processed and disposed of. The remaining inert waste is then delivered to a landfill for eventual disposal, while the organic material is routed to several processing methods like composting, biomethanation, incineration, pyrolysis, and so forth. Any local government must do an evaluation of the waste before developing a plan for a municipal solid waste management system. Waste stream assessment delivers necessary information about the waste quantity generated by humans, the composition of the waste, and its sourcs. The identification of the equipment and the eation of plans and programmes for the management system will benefit from the information periodically acquired from such assessments. Regional differences exist in waste volume and content, as well as lifestyle and socioeconomic status of the populace (Massoud, Tarhini, & Nasr, 2009). Degradable organic material and paper make up the majority of waste components, followed by inorganic material, which includes metals, plastics, textiles, glass, and other materials. The amount of waste generated may vary depending on the socioeconomic position of the neighborhood, but the fraction that ultimately makes it to urban disposal facilities stays the same. Waste differs in composition depending on the socioeconomic standing of the community; for example, biodegradable and organic waste make up a bigger portion of waste in developing nations than inorganic waste does in wealthy nations.

3.5 PROCESSING

Density, moisture content, and waste dispersion are the main causes of variation in waste composition. Typically, as waste is moved from its source to a disposal site, its density varies. The density is affected by a variety of factors during the transfer, including handling techniques, waste storage at the primary and secondary locations, exposure to the elements, decomposition of the waste, and more. Physical and chemical wastes are separated into categories. Finding the right processinends greatly on waste characterisationcharacterization. In order to develop processing procedures, choose waste handling equipment, and dispose of waste ultimately, it is required to analyze the physical properties of waste. The physical qualities are primarily determined by some factors. The moisture level is the ratio of water weight to the total weight of wet waste. Waste that has a lot of moisture interferes with a number of processing techniques that need dry waste. The weight of the waste grows along with the moisture content, which complicates collection and transportation. The economic viability of mass burning or incineration technology is determined by moisture content. An excessive quantity of energy is needed to eliminate extra moisture and to increase the temperature of the water vapor. Waste density is the mass per unit volume. The most important physical property of waste is density. To fit more waste into a sanitary landfill, it shoulde compacted to its maximum density.

Figure 3.2. Waste compaction equipment. (Source: By Ropable - Own work, Public Domain, https://common/918/compactor-landfill-grader-trash.jpg, Creative Commons License).

Any compaction equipment (Figure 3.2) can lower the volume of waste for up to 75% of it. As a result, the initial density rises by 100–400 kg/m^3. The density of waste frequently fluctuates as it passes through numerous processes, including breakdown and temperature variations, from the point of origin through final disposal.

When choosing machinery like separators and shredders that are appropriate for processing specific kinds of waste, the size of the waste is a key factor. Similar to how it is done for soil particles, size distribution analysis is carried out. Chemical properties reveal the distinct chemical components of waste as well as their potential for heating. The lipids are the fats, greases, and oils produced by food waste, cooking fats, and other sources. Since lipids have a calorific value of roughly 38,000 kJ/kg, they are better suited for energy recovery. They are biodegradable in nature and become liquid at room temperature, but the pace of decomposition is slowed down by the low solubility of lipids in water. Proteases are made up of substances like amine-containing organic waste, carbon, nitrogen, oxygen, and hydrogen. Food and garden waste are sources of proteins. On partial breakdown, proteins with the amine group release an offensive odor. Organic waste, which includes food waste and contains carbohydrates in the form of sugar polymers, makes up a significant component of MSW (Hafid et al, 2017). When in contact with oxygen, carbohydrates quickly break down and release CO2, H2O, and CH4. Combustible solid waste is analyzed by looking at the percentages of carbon, nitrogen, hydrogen, oxygen, sulfur, and ash. Halogens are also taken into account in the analysis since they influence emissions during combustion. A comprehensive examination is also carried out to determine the chemical properties of organic waste in MSW. The cellulose, proteins, lignins, lipids, hydrocarbon polymers, and inorganic content are all examined in MSW. The final analysis also looks at waste mixtures that can produce the necessary carbon-to-nitrogen ratio for biological conversion processes.

3.6 COMBUSTION

The quantity of heat produced during the combustion of solid waste under typical circumstances is known as the heating value. A solid waste sample is examined to see if it might be used as fuel for the incineration process. Waste made of plastic has a high heat value, little ash, and little to no moisture. Due to their average carbon content and moderate moisture level, paper and cardboard have intermediate heating values, but food waste or green waste has relatively low heating values because of their high moisture contents. According to numerous studies, solid waste produced in high-income countries has higher heating values because the waste contains more carbon, whereas low-income or developing countries have minimal heating value since the waste contains more moisture and less carbon. The management of waste from the point of generation, waste handling, sorting, storage, and processing at the source, along with transfer, transport, and disposal, are all parts of the functional system of MSW. Management practices weren't as important in the past, but in more recent years, doubts about public health and aesthetics have made them more important. The MSWM originally came to light about 500 BC, when there were limited solid waste management methods and procedures available (Arena, 2012). People eventually came to understand that this is one of the crucial areas of waste management that requires substantial attention. The ineffective managesolid waste in metropolitan areas of emergingdeveloping nations is increasingly not only an environmental issue but also a socioeconomic disadvantage.

Waste production has existed throughout history from the very beginning. Early in 1751, Corbyn Morris established the seeds for waste management in London. He argued that the general public's health was of the utmost importance and vigorously supported his proposal for uniform public administration. In order to establish a clean and sanitary metropolis, he made arrangements for the waste and scraps of the city to be routed to the River Thames. The eighteenth century saw the establishmentorganized

SWM in the city, according to recent a study. 50 years prior tthe Public Health Act, this was published.

(a) Ordinary Portland Cement (OPC)

(b) Coal ash

(c) Lawrencepur sand

(d) Quarry dust

Figure 3.3. Coal ash. (Source: Abbass W, Abbas S, Aslam F, Ahmed A, Ahmed T, Hashir A, Mamdouh A. Manufacturing of Sustainable Untreated Coal Ash Masonry Units for Structural Applications. *Materials*. 2022; ages/2021/02/26/2021-02-26-12-02-26-960x640.jpg, Creative Commons License).

Coal ash, (Figure 3.3) which made up a large amount of the waste at the time and was used extensively in the production of bricks and soil improvement, was also present. Systems for resource recovery and waste collection become possible because of this rule. The MSW system established the first framework for waste management. An unexpected and unsettling cholera outbreak rocked the middle of the nineteenth century. Then, public health emerged as a crucial issue. The disease prevention act of 1846 marked the beginning of waste management system regulation. Every home was required to dispose of its weekly waste in movable containers, according to a law enacted by the Public Health Act of 1875. The idea of a dust bin had never been used before. Similar sorts of MSWM emerged in cities in the 20th century in Europe and North America. The first US city to implement public sector waste management was New York City in 1875 (Burian, Nix, Pitt, & Durrans, 2000). When waste management first began, solid refuse was loaded onto open trucks drawn by a couple of horses. The first closed-body trucks were produced in the early 20th century, and a dumping lever mechanism was used in Britain in 1920. Hydraulic compactors were first introduced in 1938. Every human society depends on MSWM inone way or another.

3.7 NEW TECHNOLOGIES

The financial situation is so dire that it cannot afford to adopt new technologies. People in developing nations are nonetheless conscious of the pollution caused by MSW notwithstanding their progresral public in developing nations now recognises recognizes the value of putting effective management systems in place to address the SWM issue. Waste management concerns are now generally understood. The level of demand for more practical technical support in the realm of waste management reflects growern. The problem of waste management in emergingdeveloping countries cannot be resolved by merely importing expensive and cutting-edge equipment from rich nations. The creation of appropriate technology that is more useful in developing nations is urgently needed. In other instances, the adoption of expensive, complex procedures from wealthy nations has failed because those techniques were not designed for the existing management system. Municipal corporations are responsible for collecting MSW. In the megacities, house-to-house collecting is being implemented. Municipal corporations have partnered with nongovernmental organizations to manage a substantial amount of waste in cities. Private contractors have been used to carry waste from open collection sites to the disposal site or to transport waste from community bins. The collection, classification, and disposal of waste are currently under the control of various NGOs and citizen committees. In some locations, waste pickup is organized by welfare societies for a monthly fee. A certain region is designated for sweepers. In developing nations, waste collection is occasionally neglected since the streets are chaotic and packed, which is uncomfortable for residents. The majority of communities lack the resources to offer a specialized waste collection system to every area of the city. Many wealthy nations have tested techniques like WtE with good WtE is a relatively new notion in certain emergingdeveloping nations, like India. As a result, the landfill continues to be the preferred method of waste disposal. As a result of improper MSW management, 90% of waste is dumped directly into the soil of different cities and municipalities (Sharholy, Ahmad, Mahmood, & Trivedi, 2008). Such dumping causes heavy metals to leak in numerous coastal locations. In poor nations, there is typically little land that can be used for waste disposal. Without adhering to the sanitary landfill purpose, which calls for waste to be dumped at a specific location and allowed to deteriorate biologically, chemically, and physically in isolation from human settlement, MSW is typically disposed of in low-lying areas, generally outside of metropolitan centers. Incineration is a thermal treatment technique that can minimizvolume and prevent it from encroaching on the propertyland. It reduces the added environmental burden. Due to complicated technology, this method is highly expensive. Large-scale burning is required for incineration, and it also produces air pollution, which requires further management. In underdeveloped nations like India, whose waste has a high organic and moisture content and a low calorific value of 800-1100 kcal/kg, incineration is not a particularly feasible option. Anaerobic digestion and biomethanation may therefore be used in underdeveloped nations where waste has similar qualities. These less expensive methods are suitable for the process of converting rich organic matter into a renewable resource.

3.8 DUMPING

Open dumping of MSW is not recommended for the protection of the environment or human health in developing nations. 90 percent of India, 85 percent of Sri Lanka, 65 percent of Thailand, and 50 percent of China have open dumping. To establish a proper waste management system, appropriate technologies such as composting and WtE processes should be implemented. Any waste management method's viability is influenced by expenses and a region's financial standing. In industrialized nations, where the financial standing allows for the purchase of any expensive equipment, there is a systematic system of segregation and collecting; nevertheless, in developing nations, there is no such financial base. In poor nations, manual segregation and inexpensive technologies are still in use. In poor nations, proper waste segregation and recycling procedures should be followed since a lack of segregation practices results in the loss of the majority of valuable materials that can be utilized again as secondary resources. The elimination of landfill in MSWM systems is a goal of developing nations. In the 20th century, these nations grew quickly. Regular system follow-up offers accurate data for efficiently managed waste management activities, which forms the basis for technical training in these sectors. To relieve the pressure on landfills, recycling is increasing in wealthy nations. To ensure the safe disposal of waste, extended producer responsibility is being introduced. The necessary procedures can be used in place of manual work. So that processing can continue without interruption, sustained planning and funds are made available. There is a lot of knowledge concerning waste management. The present global trend supports the idea of sustainability. In Taiwan, South Korea, and Japan, national projects have begun.

An investigative idea in waste prevention is the Integrated Municipal Solid Waste Management system. It covers all facets of waste management, including programs for recycling, composting, and disposal. In wealthy nations, this idea is extensively used. System analysis is another crucial tool for managing MSW using a variety of holistic approaches. It contains specific technical models and evaluation methods for overcoming the difficulties and obstacles related to waste management. The conventional open dumping approach is still widely used on the Asian and African continents, which results in untreated MSWs and environmental degradation. In contrast, the open dumping practice is completely being phased out in North America. Most high-income countries, as well as Europe and Asia, use incineration techniques. Before beginning any implementation techniques for MSWM systems, a number of rules and processes must be followed. However, if these recommendations are not being followed in any one sector, a number of issues could prevent their adoption. The nations that are greatest at managing their implementation strategies are those that can meet environmental, social, economic, and technological goals. Because of the numerous issues they encounter in various sectors, implementing MSW management is difficult in developing nations. Rapid urbanization and population expansion in developing nations can often exceed the capacity of the system, making it difficult to manage waste (Marshall & Farahbakhsh, 2013).

Waste management, the environment, and public health are all related. Waste management hierarchy is one of the biggest challenges and it requires a wide and comprehensive range of diverse treatment options, such as composting and recycling, to form a reliable infrastructure. If proper waste management is not followed, it affects both public health and the environment, causing a major disruption in the waste management system. Despite effective planning models, an important factor in determining the amount of material removed from the collected waste is the lack of knowledge regarding segregation at source. The nature and amount of waste are changing in developing nations due to increased urbanization and bettering lifestyles. Municipal corporations are utterly incapable of finishing the job. More than 25% of the garbage is still not collected. According to recent research, improper waste management poses major risks to human health. Developing nations do a good job of solving this issue. Practically speaking, municipal administrations typically allot small expenditures to dealing with waste. Less money is made available since the public is less aware of the problems with waste management, which ultimately results in a further decline in service quality. Not much of the money raised from the SWM is used specifically for waste management; instead, it is put into general municipal funds. When money is given to the federal government rather than local governments, the misallocation of waste revenues is made worse. The amount of revenue received and the actual degree of service provided are unrelated. Due to a lack of knowledge and the accessibility of conventional waste management methods in developing nations, MSWM encounters numerous issues. There issize-fits-all solution to the solid waste issue in emergingdeveloping nations.

3.9 POPULATION AND WASTE

The majority of people on Earth reside in developing nations. Given that these populations are growing more quickly than they ever have in industrialized nations, waste management in these nations is a pressing issue that needs to be addressed. To develop and run the functional components involved in solid waste management, it is essential to understand the source of municipal solid waste generation. It is vital to gather data regarding the source and rate of waste generation in a region in order to evaluate the present and future requirements for the budgeting, operating, and processing of MSW. Depending on a variety of variables, including economic background, population density, industrialization, current management status, and lifestyle, the rate of solid waste generation varies from community to community, city to city, and country to country. Confusion has been caused by the persistence of MSW classifications, which also make it challenging to analyze and interpret the outcomes of different sources of generation. In the majority of developing nations, households are the main source of MSW generation. It is also referred to as domestic or residential waste and is produced by daily household usage. Examples include food waste, paper, plastic, metal, and glass containers, dirty papers, and more. Nearly two-thirds of all waste in underdeveloped nations comes from the kitchen (Abd El-Salam, 2010)collection procedures, the type of temporary storage containers, on-site transport and central storage area, treatment of wastes, off-site transport, and final disposal options. This study indicated that the quantity of medical waste generated by these hospitals was 1.249tons/day. Almost two-thirds was waste similar to domestic waste. The remainder (38.9%.

Developing nations have a considerable increase in institutions including schools, government buildings, and hospitals along with population expansion. In most institutions, paper makes up the majority of the waste category. Paper, waste, litter, sand, stones, spilled loads, and wreckage from auto accidents make up this mess. Lack of public awareness, littering by vehicles, roadside occupants, and pedestrians are the main causes of waste entering the streets. The primary driver of MSW creation on the streets is public education in relation to participation. Waste on the streets has a negative visual impact, especially on visitors, and thus has an indirect impact on a city's economy.

Urbanizatiodustrialization trends have caused a large amount of emergingdeveloping nations to produce waste. Broken concrete, scrap and wood products, board, glass, outdated electrical equipment, tilling and related brickwork, metal, and paints are typical components of this sort of waste. In the majority of developing nations, agricultural waste output is one of the greatest organic sources of waste. Crops and other harvests are included in the waste produced by agricultural sources. The country's climatic conditions, which are crucial for crop development, determine the rate of waste generation from the agricultural industry. An important element for the effective implementation of waste management solutions in a nation is estimating the volume and composition of MSW generation. The amount of waste produced is influenced by a number of variables, including population growth, economic prosperity, cultural norms, geographic distribution, and commercial activity.

3.10 URBANIZATION

Of all the variables, population growth has the biggest impact on how quickly MSW is produced. In developing nations, urbanization is growing in tandem with population growth. Waste output has increased tenfold in India over the past century as a result of population growth and urban expansion. It has been shown that while roughly 45 percent of the population uses a functional sanitation system, only about 20 percent hpriate access to water. In the end, the governments of emergingdeveloping nations are confronted with significant issues that are impeding the achievement of sustainable development and sound municipal services. In developing nations, this is inversely correlated with the rate at which MSW is produced. Urbanization and, ultimately, an increase in the pace of waste generation have been caused by rapid population expansion, changing lifestyles, the need for a standard of living, and a propensity to move from stagnant and low-paying rural employment to higher-paying urban occupations. As rural populations have been drifting to urban regions in search of sources of livelihood,ialization has completely changed the infrastructure of emergingdeveloping countries. This finally resulted in urbanization and the production of MSW. The production of MSW from community activities is influenced by lifestyle, culture, and economic status.

It has been discovered that the qualities and composition of waste collected from various communities vary greatly depending on all of these parameters. For instance, some societies utilize and dispose of paper and plastic in enormous quantities, while others discard organic materials. Since MSW is a natural byproduct of production and consuming activities in any economy, it is frequently asserted that its generation is correlated with economic activity and the degree of production in various sectors of a society. For waste categorization and evaluation as well as choosing the best waste management system to meet the conditions and requirements at hand, the proper information pertaining to the existing sectors is crucial. Cities with low and intermediate incomes produce a significant amount of organic waste, whereas those with high incomes produce a more diverse mix of waste, including higher percentages of paper and plastic. It has been difficult to address the enormous volumes of dailaste produced because of people's ignorance of and indifference in to waste management. Particularly in developing nations, where severe issues like illiteracy, poverty, and inconsiderate waste management exist, there is a dilemma. People frequently believe that waste management or collection is the municipality's job, making them free from any accountability.

They consequently don't give waste problems much thought, and as a result, they rarely try to dispose of their waste properly. This significantly affects the global pace of MSW formation. The pace of MSW generation is also impacted by the availability of fruits, vegetables, and annual crops. For instance, during the summer, the quick deterioration of a waste component affects the pace of waste creation. Open dumping and burning of MSW pose a serious hazard to the environment and all living things in developing nations. In developing nations, the production of MSW is growing steadily at a rate of 3–7% annually (Subramanian, 2000). The main problems with MSW generation in developing nations are brought on by substantial population growth in urban areas, a lack of legislation and policies for realistic long-term planning, inadequate storage, limited collection, improper disposal, a lack of suitable technology and equipment, scavenging, a lack of skilled labor, etc.

In developing nations, paper, kitchen waste, plastics, metals, textiles, rubber, and glass make up the majority of the materials that make up MSW. The per capita generation rate in developing nations ranges from 0.4 to 1.1 kg/d, peaking at 2.4 kg/d in some urban areas and more in tourist destinations. It has been discovered that developing nations produce less waste overall and a greater percentage of organic material in the MSW stream. In underdeveloped nations, it is estimated that at least 50% of waste is organic in composition. Although they are getting better, MSW services in developing nations still lag behind those in industrialized nations. Although it is still in its infancy, municid waste management is predicted to become more prevalent in emergingdeveloping nations. Concerns with MSWM have become more important as a result of efforts to lessen the drastic disposal of MSW to landfills in developing nations. According to a 2012 World Bank assessmentst Asia and Pacific region would produce more than 40% of the world's global MSW by 2025, with China alone creating almost 180 million tonnes of MSW. The tremendous pace of MSW generation in developing natis directly correlated with a country's progress and development. Lack The lack of a management system causes more environmental degradation and increases health concerns for urban residents. Ghana has a reliable door-to-door garbage collection system, which is typical in most densely populated places. The Ghana government has enacted the EPA and ensured that waste is collected, disposed of, and treated in a sustainable manner by all parties, including the business sector. Additionally, it strengthens local level planning, oversight, and law enforcement. Ghana, however, lacks strict legislation for the handling of solid waste.

Sri Lanka is having MSWM issues as a result of its climate (Vidanaarachchi, Yuen, & Pilapitiya, 2006). As ragpickers move from street to street and visit waste yards in search of precious items, they are susceptible to a number of ailments, including skin conditions. Increased per capita municipal waste production in developing nations is a result of rapid urbanization, population growth, and lifestyle changes. In underdeveloped nations, MSW is collected manually or mechanically. Waste collection is complicated because MSW includes waste from both specific locations and unidentified sources. The waste is kept in the designated bins. Waste is picked up from door to door or by motorized trucks that arrive at intervals. Waste collection is entirely the responsibility of the local governments. Either through their own infrastructures or through agreements with the commercial sector, they must collect the waste. MSW is transported using refuse vans. Compactors, tippers, dumpers, and stationary compactors are the most common types of transportation vehicles. Sweepers and sanitation personnel either hand load and unload waste vans, or they employ mechanized processes. Transport stations are found in several cities and towns, supporting the intermediate transfer of waste from the neighborhood. The waste is either processed for energy or disposed of in a landfill. In Pakistan, the daily production rate of MSW is 2.4% annually. It is important to dispose of waste properly. Sometimes, low-lying places and aquatic bodies are used to dispose of the waste. Open dumping is the most popular method of treatment in Pakistan; incinerators and sanitary landfills are both relatively recent. MSW is among the categories of solid waste in Sri Lanka that is causing increasing concern. The most popular methods of waste management are open burning and unreliable landfilling.

3.11 REGULATIONS

The Governmof India created the Municipal Solid Waste Rules; revisions are in the progress making it mandatory for urban local bodies to improve their solid waste management systems as envisaged in the rules. This was done in response to the serious environmental degradation caused by the unscientific disposal of MSW. These regulations specify how waste should be collected, separated, stored, transported, processed, and disposed of. The regulations also require all cities to provide appropriate waste treatment and disposal facilities. Standards for compost quality and safety are also outlined in these regulations. The MSW guidelines also specify compliance standards for each and every stage of waste management, including collection, source-segregation, transportation, processing, and final disposal. MSW littering is forbidden in all cities, towns, and metropolitan areas that have been alerted by the state governments (Kumar et al., 2016). The municipal authority must take actions to prevent littering and encourage compliance. Arranging for house-to-house MSW collection using any of the following strategies: community bin collection, house-to-house collection, collection on regular, informed timings, and scheduling by using bell ringing of musical vehicles, etc. is an important practice. We need to come up with a plan for collecting waste from slum communities, including hotels, eateries, oice buildings, and commercial districts. Biodegradable wastes from slaughterer shops, fish and meat markets, and fruit and vegetable markets must be handled to make use of them. Biomedical and industrial wastes should not be combined with MSWs and should adhere to the rules that are specifically outlined for them. Hand-driven container carts or other small vehicles must be used to transport collected garbage from residential and other places to the community bin. Horticultural and construction or demolition wastes or debris must be collected and disposed of separately in accordance with the appropriate standards.

Stray animals must be maintained in accordance with State legislation and cannot be permitted to roam around landfills or any other area of the city or municipality. For the benefit of the general public in a city or town, the municipal body shall inform the waste collection schedulend the likely manner to be used. Municipal authorities must establish programmes to raise awareness of waste segregation and encourage recycling or reuse of segregated materials in order to motivate citizens. To guarantee community involvemc in waste segregation, the municipal authority must implement a staged programme. The municipal authorities shall schedule frequent meetings with representatives of neighborhood resident welfare associations and non-governmental organizations at quarterly intervals for this purpose. Municipal authorities must build and manage storage facilities such that the area around them doesn't become unsanitary. In order to construct and establish storage facilities, it is necessary to consider both the population density and the amount of waste produced in a particular location. A storage facility must be positioned so that users can access it. Wastes held in facilities set up by municipal authorities or any other agency must not be exposed to the open air, and they must also be visually pleasing and user-friendly.

Waste processing, transfer, and transportation must be made simple by the design of storage bins or facilities (Kumar et al., 2009)which are facing a serious pollution problem due to the generation of huge quantities of solid waste. This paper presents an assessment of the existing situation of municipal solid waste management (MSWM. Bins for biodegradable waste storage must be painted green, recyclable waste storage bins must be printed white, and other waste storage bins must be printed black. Waste handling by hand is not permitted. Manual handling must be done carefully and with sufficient precautions for worker safety if it is necessary due to limits. Waste hauling vehicles must have covers on them. Waste should not be left out in the open or visible to the public to avoid spreading. The garbage must be removed daily from the storage facilities set up by local authorities. Before they begin to overflow, the bins or containers wherever they are located must be cleaned. Waste handling several times before final disposal shall be prevented by the design of transportation vehicles. Municipal authorities must utilize appropriate technology, or a mix of technologies, to handle waste in order to reduce the strain on landfills.

Figure 3.4. Biodegradable wastes. (Source: https://upload.wikimedia.org/wikipedia/commons/a/a1/Biodegradable_waste_container_3.jpg, Creative Commons License).

Biodegradable wastes (Figure 3.4) must be stabilized through composting, vermicomposting, anaerobic digestion, or another suitable biological processing. Mixed waste that contains reusable materials must go through the recycling process. Pelletization, along with incineration with or without energy recovery, is another option for processing waste in some circumstances. Before requesting permission to utilize such cutting-edge technology, municipal authorities or facility operators must contact the Central Pollution Control Board to have the standards established. The only waste that can be deposited in the landfill is non-biodegradable, inert waste, and other waste that cannot be recycled or processed biologically or by any methods. Land filling must be done with regard to the leftovers/ residues from waste processing plants and the pre-processing rejects from those facilities. Unless it is determined that mixed waste cannot be processed, land filling of such waste must be avoided. Land-filling must be done in accordance with proper regulations when unavoidable conditions exist or until alternative facilities are installed. The management of the civic and infrastructure requirements of any given region falls mostly under the purview of the municipal corporation. The municipal corporation is responsible for maintaining the cemeteries, water supply, sewage treatment and disposal, open local spaces, roads, streets, and flyovers. It also covers the modernization and improvement of the waste disposal system. The collection and transportation of waste is a major concern for any municipal organization and necessitates personnel as well as a variety of transport equipment, including compactors, tippers, dumpers, stationary compactors, and so forth. The enormous challenge of waste management calls for effective coordination. There are many tiers of staff deployed, starting with motor loaders and sweepers and moving up to administrators and managers.

3.12 COLLECTION OF WASTE

The municipal corporation is in charge of acquiring and standardizing waste cans (Wen et al., 2018)implementation, and evaluation of a sensor-based Internet of Things (IoT. It manages the use of nongovernmental organization workers at collection stations as well. Municipal corporations are being automated to improve the efficiency of the system, and their responsibilities are expanding to include managing electronic systems like vehicle monitoring and the use of radio frequency identification. According to the debris management policy, the municipal corporation also strives to implement the disposal of construction and demolition waste. The municipal corporation also raises public awareness by staging various plays and activities. It plays a critical role in introducing the newest technologies and changing corporate culture. Along with the formal sector of MSWM, there is also an informal and unofficial sector made up of individuals who work there. The workers in this industry, known as ragpickers, are highly susceptible and pick up rags as well as recyclable debris and other resources. They perform labor-intensive, low-tech, low-paying, unrecorded, unregulated work that is frequently carried out by individuals or family groups, and their operations are the epitome of the informal sector. The movement of MSW from the point of generation to the point of treatment or disposal is referred to as the waste collecting process. The main goal of MSW collection is to move MSW from the source to a location where the loaded vehicle can be discarded, in addition to collecting recyclable solid waste products. This location could be referred to as a transfer station, landfill, or waste processing plant. To prevent the environmental effects of MSW on urban residents, this location must be far from populated areas. Every municipal organization is required to periodically notify the public of the waste collection schedule. The segregation and transportation of waste to authorized collectors in accordance with the municipal corporation's timetable should be the responsibility of the solid waste generator. The municipal authority should frequently launch awareness drives in collaboration with nearby towns and nonprofit organizations to educate individuals about proper waste segregation and collection. Users must pay for this form of waste collection system where waste collectors visit each home to collect waste.

3.13 STORAGE

The short-term storage of waste in enclosed containers rarely causes any issues in houses with reasonably large open areas. The majority of homes need waste disposal machines because there isn't much room for waste containers in the kitchen's working area in tiny flats, which limits storage to a 24-hour period. As a result, either daily waste collection from each residence is required, or a decentralized communal waste container is provided. For proper management, regular waste collection from each unit is required, whereas market stalls may require collection services three to four times per day. The population density of most parts of big cities is significantly higher in developing nations like India than it is in developed nations. In these heavily inhabited locations, there aren't many external sites for waste storage, hence the collection frequency should be four times daily. This kind of bin system is implemented in areas where locals can conveniently access waste cans to dispose of their waste. Containers need to be covered and cleaned frequently in order to use this method effectively. Additionally, separate bins for paper, biodegradables, and recyclables should be available. Community waste cans haven't always been adequately maintained or received by residents, which has led to a majority of cans not being emptied on time and eventually overflowing, producing unsanitary conditions in the neighborhood. Citizens frequently toss waste into collection bins far away because they don't like the smells coming from the containers, which exacerbates the issue. Community bin collectors can still be utilized in some circumstances, though, and they are highly effective at collecting waste. For instance, it is possible to expand the number of waste cans and put them in high-rise structures, housing complexes, and small communal places. However, regular collection and cleaning are necessary to prevent unpleasant odors brought on by anaerobic waste degradation and to increase community acceptability of waste cans. Community waste cans, however, need to be well-designed for convenient access by locals. Otherwise, depending on the type of equipment the municipal or local organization uses, the containers can be mechanically unloaded off vehicles. The major phases in final disposal include handling and sorting waste.

Moving loaded containers to the location of collection is considered handling. A crucial step in the collection, management, and safe storage of solid waste at the source level is waste segregation. Two essential processes are involved in transportation: first, the waste must be moved from small collecting vehicles to large trucks, and then it must be moved to a landfill site remote from the metropolitan area for final disposal. In the management of solid waste, MSW collection speed is crucial. This can be accomplished by better vehicle selection in both urban and rural settings, taking into account the types and amounts of waste, the state of the roads, and the distance to the final disposal site. For medium shipping routes of less than 10 km to the dumping site, tractor-trailers are most suited. For trips greater than 10 km, trucks are the most practical option due to the associated costs. The collection of low density waste like paper, plastics, and packaging material, less than 250 kg/m^{3}ich may be the most prevalent waste in urban regions and cities is best suited for compactor trucks (Y. Wang et al., 2010)but close to that of low-density (expanded polystyrene. Due to the high density of the waste, namely, dust, dirt, and organic waste, compactor trucks are not appropriate for rural areas or small towns. Utilizing a variety of vehicles, the MSW collected from waste bins is delivered to the final disposal locations. Tricycles or tractor-trailers are frequently used for the delivery of garbage in cities and rural areas. Large towns and cities are where you'll mostly find lorries and light motor vehicles. Trucks are utilized in metropolitan areas to move garbage from collection sites to disposal locations.

Transfer stations are places where MSWs are temporarily stored after being unloaded from collection vehicles before being reloaded onto large vehicles for long-distance transit to final disposal locations or landfills.

By merging the cargoes of multiple different waste collection trucks into a single load, communities can save money on the labor and operation costs of transporting waste to distant locations. Consequently, there are fewer vehicles driving to and from the disposal facility overall. Waste transfer stations offer the benefit of lessening the traffic impact of trucks traveling to and from the dump site, but they can also increase traffic in the area around where they are located. To prevent issues for surrounding homes, transfer stations should be carefully situated, built, and run. To create a strategy that is both economical and environmentally sustainable, an integrated MSW system is required. The majority of spending on solid waste management is acknowledged to be accounted for by municipal garbage collection on a global scale. The collection of waste accounts for 80–90% and 50–80% of the budgets for waste management in low- and middle-income nations, respectively. Therefore, the establishment of an integrated waste management system is hampered by problems with MSW collection and transportation. The development of sound solutions by civil authorities is encouraged, particularly in metropolitan areas where transportation and collecting costs are high. As a result, a key consideration in the design of a waste management system is the optimization of MSW collection and transport from the source. Waste sources are dispeosition of the starting materials, promotes the rate of MSW breakdown and the activity of these microorganisms.

3.14 MICROORGANISMS FOR WASTE DEGRADATION

The process of composting is the controlled microbial decomposition of solid waste's organic component under aerobic circumstances when microorganisms turn waste into a stable end product like compost. The composting of two or more materials simultaneously is referred to as co-composting. In the composting process, the organic portion of waste decomposes, resulting in a drop in volume, weight, and moisture content, a reduction in potential odor, a reduction in pathogens, and an increase in potential nutrients for agricultural use. Both industrialized and developing nations have used this method as a contemporary alternative to traditional waste management for many years. It improves the economic and environmental sustainability of waste management systems by diverting a sizable amount of organic waste from municipal collection services and from final disposal locations. Organic waste that degrades quickly stabilizes into carbon dioxide and water during composting. Users cannot completely stabilize a by-product during the composting process since doing so would impair the compost's ability to produce soil. Instead, stabilization depends on how the by-product is used by the final consumer. High temperatures speed up the breakdown of proteins, lipids, and complex carbohydrates like cellulose and hemicellulose during the thermophilic phase. When the compost reaches its final stage of maturation, mesophilic microbes once again flourish over other types of microorganisms when these high-energy molecules progressively run out. The management of the carbon-to-nitrogen ratio, pH, temperature, moisture content, and other nutrient levels, as well as by the composition of the starting materials, promotes the rate of MSW breakdown and the activity of these microorganisms.

Microorganisms are crucial to the breakdown of organic waste (Mupambwa & Mnkeni, 2018). As they develop and reproduce, numerous kinds of invisible aerobic microbes break down organic matter. The nutrient balance, moisture content, oxygen supply, temperature, and pH levels must all be kept in check for the composting process to go properly. All the factors must be balanced and managed in order to speed up the composting process. In the presence of diverse microbes, complex stuff decomposes into a simpler form during the chemical changes in composting. The microbes need enough energy for these operations before they can manufacture new cellular material. These heterotrophic microbes use fermentation and respiration as their primary energy routes. Both aerobic and anaerobic respiration are possible. Because it creates a lot of energy and boosts temperature during the process, aerobic respiration is more efficient than anaerobic respiration. Additionally, it produces less odor since it is more efficient than anaerobic respiration. Due to the synthesis of carbon dioxide and water from carbonaceous material during aerobic composting, a significant quantity of energy is released. Each intermediate chemical serves as the catalyst for more reactions during the course of this process, which is carried out by a succession of reactions. In anaerobic respiration, the microbes receive energy from sources other than oxygen. Carbonates, sulfates, and nitrates are some of these electron acceptors. They utilize these alternative electron acceptors in their energy-producing metabolism, which results in the production of methane and hydrogen sulfide and other undesirable gases. The functioning of aerobic microorganisms is impacted by organic acid intermediates produced during anaerobic respiration, whereas in aerobic respiration these intermediates are quickly consumed in subsequent reactions and do not cause odor as in anaerobic respiration. Regardless of whether the C:N proportion, wetness, or air circulation fall outside the guidelines for fertilizing the soil, normal deterioration happens in any pile of waste material.

Deterioration keeps reintroducing waste at a rate that is immediately noticeable and may also produce unpleasant odors. Fertilizing the soil is a deliberate effort to speed up general decomposition, reduce odor production, destroy viruses, etc. In order to achieve a high rate of microbial movement, the fertilizer blend must be designed to optimize conditions inside the heap in terms of nutrition, the presence of oxygen and moisture, as well as pH and temperature levels. The primary substrate is the waste that is composted, whereas the amendment is the substance that is added to the waste sample to balance its pH, C:N ratio, and moisture level. Bulking agents are added to the mixture to preserve the compost pile's structure and porosity. These elements keep the compost pile's physical and nutritional qualities intact, fostering the ideal environment for optimum microbial activity. The two most important properties of a compost mix that must be maintained in order to have quick composting are the C:N ratio and moisture content. An amendment is added if any one of the qualities is not met.

CHAPTER 4

Waste Crisis

CONTENTS

4.1 INTRODUCTION

On Earth, we waste a lot of resources. Although it is not our intention, this is an ingrained and unavoidable aspect of human society. Sawdust (Figure 4.1) from sawing wood, metal flakes from drilling and soldering circuit boards, sludge from chemical reactions, restaurant leftovers, tonnes of waste paper from court cases and hearings on the environment, dirty diapers, and other home waste are all there.

Figure 4.1. Sawdust. (Source: https://upload.wikimedia.org/wikipedia/commons/6/6b/Sawdust..jpg, Creative Commons License).

The waste produced by each family to the highly harmful industrial wastes from the making of specialized goods like electronics, computers, autos, petrochemicals, and plastics are all examples of societal wastes. Waste is produced in almost every facet of our daily lives. Waste is unavoidable. In the last ten years, there has been an increasing understanding that the resources on our planet are limited and that the ecosystem is being steadily but silently destroyed. Currently, about 20% of municipal waste in North America is recycled; for example, glass and plastic are sent to companies to be made into new items, while paper and metal cans are returned to pulp mills (Greco et lni, 2015). As recycling initiatives continue to grow, it is projected that eventually up to 50%, and possibly even more, of all residential and business waste will be recycled.

4.2 WASTE ISSUES

A portion of the waste is burned. This can be seen as a sort of recycling, the transformation of waste into energy, a very useful product when it is combined with the production of electricity, useful steam, or heat. It aids in the preservation of priceless non-renewable resources like gas, oil, and coal. However, a lot of people are worried about the ash that is created and the emissions that are released into the air. In the United States, approximately 18 percent of municipal solid waste is now burned, with approximately 75 percent of the incinerators producing electricity. In Canadar less than 5% of municipal solid waste is burned, compared to around 34% in Japan. In North America, there are tens of thousands of landfills that serve acesites for the waste that we produce on a daily basis. Landfills continue to be the backbone of waste management, despite current efforts to minimize waste levels through recycling and incineration. People relied on landfills for so long that they just took them for granted as the best option to dispose of waste. Urbanization and population expansion are strongly intertwined, as is the issue of waste disposal. It wouldn't be difficult to use a small portion of the land for waste disposal if every family still had access to vast plots of land. But in today's urbanized world, hundreds of households are frequently packed onto a single acre, making the situation extremely different.

The sheer quantity of landfills and the vast areas of valuable land they occupyre a serious issues. Every city and municipahave has at least one active or closed disposal site, however, there has been a move toward fewer but much bigger regional landfills in recent years. Landfills have long been a crucial component of commercial and industrial operations. Even nations with what seems like an endless supply of vast, open landscapes cannot afford to waste territory/land on landfills. Prime agricultural land only makes up roughly 0.8% of Canada's total land area, despite the fact that it has the second-largest land mass among all nations. Furthermore, cropland within 80 kilometers of 22 major urban areas accounts for 43% of the value of agricultural production in Canada. Therefore, landfills and good agricultural land are in direct competition with each other because both are located close to the municipalities they service and are both in short supply. After landfills are shut down, no structures will be able to be built on top of them due to the explosive methane gas that is produced during decomposition. This disallows the establishment of cities, businesses, or industries. The landfills' contamination prevents these sites from being used for cultivation. We, our children, and many generations to come won't have access to this land, which is a highly valuable resource. Since almost all of these landfills are located inside or close to urban areas, where growth pressure is highest, the situation is made worse. Surface landfills' inefficient use of land runs counter to the idea of sustainable development. It is important to look for innovative waste disposal techniques that preserve precious land resources.

As the weight from above presses down on the lowest portions of the waste, settlement occurs. The dump sinks, rots, and eventually disintegrates. Rainwater that seeps into the ground leaches heavy metals from old batteries and other waste, generating a contaminated liquid that eventually seeps downhill and taints the groundwater beneath. According to the US EPA, 75% of the 75,000 landfills in the country are leaking (Read, Hudgins, & Phillips, 2001)whereby anaerobic conditions are created within the landfilled waste. Under anaerobic conditions within a landfill site slow stabilization of the waste mass occurs, producing methane, (an explosive 'green house' gas. Because landfills include materials that are far from safe, damage to the groundwater is frequently very serious and typically irreversible. There is an incredible amount of waste, especially dangerous ones. Chemicals including PCBs, lead, solvents, dioxins, DDT, benzene, CFCs, furans, and many others are among them. Idustriesy currentl produces around 45,000 distinct chemicals, and each year, another 1,000 are introduced. Unfortunately, a lot of these chemicals are hazardous, meaning they can harm other components of the intricate and delicate environmental web in addition to poisoning humans.

Figure 4.2. Landfills. (Source: https://upload.wikimedia.org/wikipedia/commons/2/28/Landfill.jpg, Creative Commons License).

Landfills (Figure 4.2) contribute to air pollution. Gases like methane, carbon dioxide, vinyl chloride, and hydrogen sulfide are produced during the decomposition of waste and slowly leak into the air around the landfill.

On a larger scale, this adds to the greenhouse effect and global warming while also degrading the air quality in the nearby area. Even contemporary landfills with cutting-edge technologies like leachate collection systems have issues. Within a few decades following their closing, if they are not already leaking, they will most likely begin to do so. The application of contemporary technologies merely delays the inevitable. Convenience served as the driving force behind landfill siting and design in the past. "Out of sight and out of mind" was the guiding principle of landfills. This method has been employed for many years, and it would have been acceptable when there were few people on the Earth and it appeared to be infinitely big. It is now becoming widely acknowledged that near-surface landfills, the foundation of practically all waste management systems, are insufficient and are having a long-term negative impact on the ecosystem rather than being out of sight and out of mind.

4.3 DEPOSITION

In the past, waste was just thrown away in convenient places like ravines, depressions, and other places that were close to the population centers that generated the waste (Hayden & Cannon, 1983). This was a valid strategy for many years, but two advancements made this method seriously environmentally problematic. First, as the population increased dramatically, more waste was produced at a time when land was becoming a more important and limited resource. Second, as a result of the technological and consumer revolution, a great number of hazardous items were produced, mainly synthetic organic chemicals like pesticides, PCBs, paint removers, and degreasers that eventually ended up in landfills. Landfills expanded in size and contained more toxic waste than before. Instead of focusing on convenience, we should look for waste disposal techniques that don't harm the environment, consume precious resources, or constrain the availability of future resources. It's difficult to break long-standing habits. We need a revolution that throws out the status quo and depends on fresh ideas and cutting-edge technology that are in line with ideologies that respect and safeguard our environment. There is still a great deal to be done, even while the seriousness of the matter is being acknowledged and some helpful measures, like streetside recying programmes, are being introduced. Perhaps we should approach waste disposal differently. We should look for disposal technologies and procedures that safeguard the environment; additionally, these procedures must be based on fundamental ideas that the general public comprehends, accepts, and embraces. It is crucial to start with the end goals in mind while trying to redesign waste management. To serve as our road map, we require objectives and a basic set of rules. A lot of government initiatives now center around the idea of sustainable development, which has gained popularity in recent years. The United States supports sustainable development as the guiding principle for how nations should conduct their trade and industry. It is critical that waste management practices adhere to this mindset since they are a component of society's larger industrial endeavors.

The integration of social, economic, and environmental systems can be seen as the essence of sustainable development. Sustainable development, as opposed to conventional economics, maximizes the objectives of the economic and social systems. Sustainable development is defined as development that satisfies the requirements of the present without bargaining the ability of future generations to satisfy their own needs by the World Commission on Environment and Development of the United Nations. To ensure that people can feed, clothe, and house, the sustainable development idea is typically applied to the front end of the industrial cycle. The preservation of our declining supply of natural resources, including forests, minerals, oil and gas, and fish supplies, is a special concern. The United Nations World Commission places a strong emphasis on reducing the pollutants released by industrial facilities, but it makes no mention of enhancing waste disposal procedures. This blind zone is representative of the way landfills are treated in our society. Waste management and disposal must be considered as part of sustainable development at the back end of the consumer cycle. For instance, groundwater and surface waterways, some of our most fundamental and priceless natural resources, are harmed by leaking landfills. Development of this kind is not sustainable. Millions of acres of land are taken from other uses like agriculture and urban development to make room for hundreds of landfills.

4.4 SUSTAINABLE DEVELOPMENT

It is crucial that we change our strategy in order to pursue a comprehensive notion of sustainable development that takes into account both the use of resources at the front end and the disposal of waste at the back end. Future waste management practices must be based on sustainable development. We must create disposal methods that do not leave a legacy that our future generations will be held accountable for. We need to start with a fundamental truth, a fundamental principle and rationally and methodically build on it to produce a specific action plan, much as mathematicians develop a mathematical proof. Although these fundamental ideas seem nearly self-evident, they offer strong direction for the practical design and advancement of waste management systems. Waste management and disposal must be done in a way that doesn't endanger the environment or human health in the present or the future. This principle imposes significant restrictions on the location and layout of disposal facilities as well as the type of waste. For instance, it basically mandates that landfill gas emissions into the sky and leachate leaks into groundwater be stopped, or at the very least lowered to levels that the ecosystem can safely absorb. The use of this idea won't be simple because a huge percentage of landfills either leak leachate or emit gases in an uncontrolled manner now or in the near future. We'll need to consider drastically different approaches to getting rid of our waste.

Wastes ought to be handled in a way that won't burden next generations. Future resources are safeguarded by this principle. In other words, our grandkids shouldn't have to invest their time and energy in cleaning up the waste our generation left behind. The definition of sustainable development as stated by the U.N. can be summed up in this notion. The greatest amount of conservation should be practiced for non-renewable resources. This idea can be used in waste management in two different ways. First, non-renewable resources shouldn't be used in the handling and disposal of wastes. This principle specifically acknowledges that land is a priceless natural resource that needs to be safeguarded. Second, this concept mandates that, to the greatest extent possible, all usable materials from waste must be retrieved before disposal. Recycling waste also reduces the amount of waste that needs to be disposed of, which has a number of advantages. These concepts can then be used to create more specific rules. In contrast to the current de facto principle of convenience, the combined principles and recommendations will offer a logical framework for creating and executing waste management systems. We all produce waste, which makes it incredibly democratic for us to all participate in finding a solution. Understanding only the technical aspects is insufficient; understanding the social and political challenges is just as crucial. These are intertwined with economic concerns, and we must look for practical, cost-effective answers. Effective future waste management cannot be left to a highly competent project team working with tremendous focus and motivation but in isolation, unlike other complicated projects like the space shuttle. The public should be actively involved in the project instead. No advancement will be made without their backing, both personally and through their elected officials. The topic of waste management is faced with both problems and curiosity as a result of this interdisciplinary union of technical and social fields. There is a crisis with waste. We produce far too much waste, particularly in North America. At the sime, there are less fewer and fewer places to dispose of waste. The general public and elected officials are refusing to establish additional landfills because they understand the inherent risks associated with them.

4.5 HISTORY OF WASTE DISPOSAL

By 500 B.C., waste had to be disposed of at least 1.5 kilometers from the city walls in Athens (Wilkinson, 1982). Each household was in charge of gathering and transporting its own waste to the disposal facility. The Roman Empire was the first to develop a waste collection service. Household waste was dumped into the streets, shoveled onto horse-drawn carts, and taken to an open pit, frequently within the neighborhood. To spare the residents their odor, deceased animals were buried in pits outside of the villages. For centuries following the decline of the Roman Empire, waste disposal in Europe remained unorganized. As long as there was plenty of land and few people throughout the Dark Ages and the Renaissance, waste was simply deposited in suitable locations and left to collect dust. The belief that nature could take care of itself persisted for a significant portion of history. Land and natural resources were increasingly scarce as the population of Europe increased and became more urbanized, and the impact of waste increased. Sweepers were employed to collect waste and animal waste from city streets, put it onto carts, and transport it outside of the settlement. Throwing waste into rivers and canals was another practical method of waste removal. Urban waste management by municipalities has gained support as it was formerly seen as exclusively an individual duty. In response to the sanitary issues brought on by the second half of the nineteenth century's fast industrialization and urbanization, the modern notion of solid waste management initially arose in the United States in the 1890s. Given the lower population density and lower level of industry in North America than in Europe, waste issues took longer to develop. However, a yellow fever outbreak in Memphis, Tennessee, in 1878, which claimed more than 5,000 lives out of a population of around 40,000, brought attention to the need for appropriate sanitation and waste management.

A growing number of American cities by the turn of the 20th century had at least a basic level of solid waste collection (Guerrero, Maas, & Hogland,

2013). After being taken out of metropolitan areas, the debris was disposed of in a number of methods, including landfilling, incineration, and disposal in the water and ocean. Pig farms, where swine were fed either fresh or cooked waste, were built in several small to medium-sized municipalities. A growing number of these waste management were being recognised recognized as being insufficient. The sharpest criticism was directed at ocean and water dumping, and by 1933, ocean dumping had been declared unlawful for municipal solid waste, with the exception of industrial and commercial wastes. Land dumping was the most popular method of waste disposal from the 1880s through the 1930s, while doubts about the health consequences of sizable open dumps were already being voiced. Using layers of waste, ash, and mud, landfills were frequently constructed by reclaiming marshes close to cities throughout the 1920s. However, it's interesting to remember that source separation and recycling were quite popular methods of waste reduction in the early 1900s.

4.6 WASTE REDUCTION

Reduction was a cutting-edge approach to waste management that was established around the turn of the 20th century (Jayal, Badurdeen, Dillon, & Jawahir, 2010). In order to reduce the amount of waste and generate energy, incineration was pioneered by the British and the Germans. Over a century ago, in Nottingham, England, municipal waste was first incinerated on a regular basis. In the middle of the nineteenth century, Great Britain developed the first plant to produce power from incineration. Despite the use of other techniques, landfills remained the most popular method of waste disposal due to the low cost of land and the low-cost technologies required. Typically, landfills were built in tidal flats, swamps, and shallow depressions. There was not much technology in use in the early to mid-1900s. Land-consuming horizontal expansion occurred when landfills grew, taking up considerable amounts of space. The development of our society's current level of overconsumption was significantly influenced by the growth of advertising and the electronic media. The end effect was a sharp rise in the type and volume of consumer products, and consequently, waste. Packaging started to dominate how items were marketed, delivered, and sold, which only made the issue worse. Today, the waste stream as a whole is made up of more than one-third of packaging. In response to the Japanese seizure of the world's rubber supply during World War II, petrochemical businesses were obliged to develop a wide range of substitute goods that could be made from crude oil. Since then, the petrochemical sector has evolved rapidly, producing a wide range of novel synthetic organic compounds, such as nylon, rayon, polystyrene, polyethylene, chlorinated organics like polychlorinated biphenyls, pesticides, and a variety of other hazardous substances. The suitability of landfill sites had to be evaluated using information on waste amounts and generation rates, available cover soil, land use, transportation routes, and other factors. Facilities were to be built and designed to take advantage of the natural resources. The sanitary landfill was a start in the right direction, but there were still some fundamental problems. Groundwater contamination, gas emissions, and associated health problems were not effectively addressed. A lot of towns kept doing open dumping and burning. In the middle of the 1960s, 94 percent of all land disposal operations in the United States, according to the federal Bureau of Solid Waste Management, were insufficient in terms of air and water pollution, and insect and rodent problems.

Figure 4.3. Groundwater contamination. (Source: Amin Al Manmi DAM, Abdullah TO, Al-Jaf PM, Al-Ansari N. Soil and Groundwater Pollution As-

sessment and Delineation of Intensity Risk Map in Sulaymaniyah City, NE of Iraq. *Water*. 2019; 11(10):215s/pollution-from-the-sea-960x632.jpg, Creative Commons License).

Groundwater contamination (Figure 4.3) from landfills was becoming more widely acknowledged by the 1970s and 1980s (Siddiqua, Hahladakis, & Al-Attiya, 2022). A vital resource that can have a direct influence on human health when used as a drinking or irrigation supply is groundwater. The issue was made worse by the fact that it is very challenging to clean up contaminated groundwater once it has happened. Landfill management looked to science and engineering for answers to the groundwater issue. The fundamental landfill was largely left alone, although some elements were added. To prevent leachate from leaving the landfill, bottom liners consisting of clay or synthetic materials like impermeable high-density polyethylene were introduced. To reduce precipitation infiltration, caps constructed of related materials were put on top of the waste. To absorb leachate and gas, engineered collecting systems were also erected. Groundwater, surface water, and gas emissions monitoring have all become common in landfill operations. Despite these technological advancements, groundwater contamination remained a source of concern. Leachate leakage was noted as an issue in several landfills in the late 1970s by a number of studies. The increasing application of engineering methods would merely delay the development of groundwater contamination, not stop it. The danger that municipal landfills pose has come to light. Municipal sanitary landfills were found to have contaminants with similar concentrations and carcinogenic potential to those reported in industrial waste landfills. The sites also contained used motor oil, oil-based paints, baes, cosmetics, solvents, insecticides, tyrestires, and much more in addition to outdated newspapers. These chemicals include heavy metals, copper, and lead, as well as chlorinated organic compounds like vinyl chloride and trichloroethylene. Lead can harm the nervous and reproductive systems, while vinyl chloride, a gas, can result in liver cancer and neurological iss

The 1930s had seen a rise in popularity for of incinerators. However, new air emission regulations were imposed by the U.S. Air Quality Act of 1967, requiring operators of older incinerators to install air pollution controls including scrubbers and precipitators. This act ultimately priced incineration out of the market because they were already more expensive and capital-intensive than landfills. 100 incinerators had been shut down within five years of the act, and the majority of the remaining ones did so during the following few years. In North America, the solid waste industry underwent changes in the 1970s. A lengthy era of prosperity combined with the consumerism lifestyle resulted in the creation of hitherto unheard-of quantities of waste. A crisis mood had emerged by the 1980s. The inadequate management of municipal waste was acknowledged by the general population. Growing waste volumes, groundwater pollution from landfills, and both the cost and air pollution from incinerators were problems. The Resource Recovery and Conservation Act was adopted by the US Congress in 1976(Kohrell, 2020). The majority of the act dealt with hazardous wastes. The states were obligated to list their open dumps and either bring them up to standards or close them, while the Environmental Protection Agency was required to set rules for landfills based on adverse effects on health or the environment. Shutting down non-compliant landfills and dumps was permitted for a five-year window. This timeline resulted in a problem at the municipal level and drawn-out legal fights to find new locations or increase the capacity of current ones. By the 1980s, the issue of waste disposal had become a major national concern that affected every level of government. Numerous laws were developed to address this situation, resulting in a complicated regulatory system where federal, state, regional, and local organizations were given the power to examine and approve landfills and other waste management facilities. Between 1979 and 1993, the number of municipal landfills in the US decreased significantly, from around 20,000 to just 5,300. The strict new regulations were the main contributing factor, which was furthered by the difficulties to locate new landfills in the face of mounting public opposition to the presence of landfills in their neighborhoods. It seems sense that energy production through cremation would seem to be a desirable choice in this situation. In the middle of the 1980s, resource recovery in the form of recycling and energy recovery by incineration also gained popularity.

4.7 LANDFILL EMERGENCE

The process of building additional landfills was about to get more challenging. It was just a matter of time before landfills faced considerable criticism as metropolitan areas continued to grow quickly. In the 1980s, neighborhood and community organizations formed with the express purpose of preventing landfill construction in their communities. Their legal actions and lobbying, which profited from the abundance of rules and regulations being issued, were added to their community demonstrations. Even though they were based in the affected areas, the networks and coalitions formed included practitioners skilled in the politics of democratic mobilization. They may successfully reply to inquiries and offer support from other regions. The resistance to landfills and related waste management facilities, like incinerators, has been quite effective in either completely halting projects or significantly delaying them. Any new waste management initiative will now inevitably face fierce local opposition. This form of opposition is known as "Not in my backyard" or NIMBY. People who are interested in working in the field of solid wasent need to have a firm grasp of this phenomena attitude and know how to deal with it. As a result, existing landfills have gotten bigger because their capacity to develop horizontally has been constrained. Because many of these older landfills, which have evolved into super dumps, lack sufficient liners or are not located in areas where the environment may be protected, the outcome is frequently disastrous.

As a result of various solutions to issues, landfill design has developed. Corrective action wasn't taken until after an issue had been discovered or raised enough concern. Regulatory requirements were generally the driving force behind these improvements. Recently, rules for the collection and management of landfill gases were developed by the US Environmental Protection Agency (Powell, Townsend, & Zimmerman, 2016). There will likely be a spread of EPA requirements to other nations and jurisdictions because of the worldwide challenges of ozone layer depletion, global warming, and ground-level smog. We may anticipate that various new emission-control technologies will be created in response to these laws. The integrity of landfills over the long term is another issue that needs to be addressed in the future. Even when well-located and equipped with the most recent engineering advances, landfills close to major urban centers and at the ground level only offer temporary safety.

The long-term effects of natural degradation and the encroachment of civilization simply aren't something they can sustain. The industry is creating an integrated waste management plan rather than merely depending on landfills, as it has done since the dawn of civilization. Utilizing all available waste management techniques, particularly reduction/reuse/recycling and incineration in addition to landfills, is the goal in order to lessen the impact on the environment. In many countries, an integrated waste management approach is mandated by law, and it is now implemented in the majority of North American towns. For instance, the majority of states in the US have made recycling mandatory and set objectives to cut waste per capita by 25% to 50% over the course of 4–10 years (Memon, 2010)reuse, and recycle.

Source reduction process is to lessen the quantity of waste that is initially produced. This can be done in a number of ways, including buying products with minimal packaging, creating products that are more robust and simpler to repair, switching to reusable goods instead of single-use disposables, and enacting tax and other financial incentives to inspire producers to produce less waste and use fewer resources. Reuse refers to reusing a product as opposed to throwing it away. Refilling bottles is one example, as are fixing outdated furniture, appliances, cars, and appliances. The goal of recycling is to turn waste into something that can be used. Examples include recycling paper, glass bottles, plastic, metal, and aluminum cans, and paper. Another recycling method is composting, which turns organic waste products like yard waste into beneficial garden mulch. Since ancient times, waste has been burned in order to reduce the amount of waste that needs to be disposed of. It can be seen as a kind of recycling when combined with the capture of thermal energy fotion of power or other uses. To improve the qualities quality of wastes before disposal, they might be chemically or physically treated. Treatment can lessen the toxicity of waste. Before being placed in a landfill, for instance, incinerator ash can be embedded in concrete to increase its resistance to leaching. Even after vigorously implementing the other techniques, some waste will ultimately persist. Since there are currently no widely used alternatives for disposal, this waste will be dumped in landfills. However, the landfill is not usually the last stop on the journey. Landfills can be excavated, and their contents repurposed or burned for energy, even if this is not a typical practice. An integrated waste management system's design and implementation are challenging tasks that must take into account a variety of elements, such as market accessibility for recycled goods, demographics, and the availability of land and other resources. So that the people's opinions can be gathered and incorporated into the master plan, a procedure should be devised for communicating with the public. A variety of technologies must be chosen in order to design an integrated waste management system.

Landfills should be considered permanent structures since they will require monitoring and upkeep for hundreds of years after they are closed. Incinerators and material recycling facilities, in contrast, are transitory constructions that may be decommissioned and demolished after their useful span is over. The land can then be sold and used for other purposes. The cost of environmental degradation, however, is the most important consideration. The worth of the natural ecosystems on Earth and the services they render are now understood to be less valuable than the economic services and manufactured goods in commercial markets. Ecosystems contribute to, and are actually necessary for, uman welfare; as a result, they account for a portion of the overall economic value. Due to the infrequent inclusion of this value in economic assessments, environmental factors are not sufficiently taken into account when making policy decisions. Recycling facilities for materials are a prime illustration of how traditional economics ignores the worth of our ecosystem (Wilson, Velis, & Cheeseman, 2006). Financial studies of waste management facilities should take extra cost considerations into account in order to accurately reflect the worth of Earth's ecosyste The price of incinerators should also account for the damage of to the atmosphere caused by gas emissions. In contrast, since they don't seem to produce any noticeable long-term harm to ecosystems, material recycling facilities may not be subject to any such environmental costs. Unfortunately, because there are no widely agreed techniques, ecological values are not currently easily included in financial evaluations. It is recommended that economists conduct the required research to address this issue. Environmental aspects in waste management facilities cannot be quantified, but that does not imply they should be disregarded. It is preferable to use qualitative methods like preferences and weighting variables.

Figure 4.4. Refillable glass bottles. (Source: https://upload.wikimedia.org/wikipedia/commons/1/15/7-Up_Bottlers_Flavor_jug.jpg, Creative Commons License).

Although substantial improvements in waste diversion are being made, they mostly help to counteract the rise in waste formation brought on by population expansion. As a result, there has only been a minimal net reduction in waste sent to landfills. As a result, additional reductions will become progressively more challenging. It would be decades before we could stop relying odfills, even if incredible new technology for waste diversion were was created. The obvious conclusion is that, despite their importance, the three Rs do not currently provide a comprehensive approach to waste management. Additional techniques are needed to really reduce the volume of overall waste. Incineration is the only technology that can now make such aontribution. Additionally, even with aggressive recycling programmes and waste incineration, it will still be necessary to dump wastes into landfills, albeit in reduced amounts.

4.8 RECYCLING

In some places, mobility might even have slowed down. Refillable glass bottles (Figure 4.4) were used to package 89 percent of soft drinks and 50% of beer sales in the United States in 1964 (Busch, 1987). Refillable bottles were only used in eleven states and represented just around 7% of the market in 1993. In Europe, nonrefillable beverage containers are prohibited in Denmark, Finland, and Germany, respectively, and 95% of soft drink, beer, wine, and spirits containers are refillable. The use of shopping bags is another illustration of North American consumers' opposition to waste reduction. The question of whether plastic or paper bags are better for the environment has generated a lot of discussion. Stores in the Netherlands charge for the provision of bags. The waste dilemma in Japan was identified as a major issue early after World War II. Japacreated the world's most extensive and effective recycling programmes. These can serve as achievable goals with realistic targets. Japan is subject to many constraints because of its location. It is an island country with a modest geographical area and a sizable populace. In comparison to the United States, where there are only 50 people per square kilometer of usable terrain, almost 1,400 people reside there. Land is a scarce resource, so it shouldn't be wasted on landfills. Japan has made a concerted effort to lessen its reliance on imported raw materials by saving and recycling as much of its natural resources as it can. For instance, Japan imported 65 percent of its timber, 92 percent of its coal, and 99.8 percent of its oil between 1980 and 1985. In comparison to North American practices, conservation is significantly superior. Japan has only managed to attain recycling rates of roughly 50%, despite its admirable efforts. Furthermore, according to Japanese officials, there is only a very small chance for development to continue; they estimate that another 3 percent or so would be possible. In addition to their effective recycling programmes, the Japanese have turned to incinerators to prevent the use of their valuable land resources as landfills. They constructed 1,900 incinerators in the 25 years leading up to 1987, of which 69 percent produce electricity. They have determined that incinerators must play a crucial role in municipal waste management despite uncertainties about dioxin emissions, air quality, and emissions from old units. The reduction of waste going to landfills is Japan's main objective. We must imitate the Japanese strategy in North America. Waste needs to be viewed as a resource and important economic good.

Due to our expanding population, North America generates an increasing amount of waste. Recycling cannot possibly keep up with this increasing waste pile. Incineration is the only technology on the market right now that significantly reduces the amount of waste that ends up in landfills. Recycling and incineration can be combined to achieve the overall objective of reducing our reliance on landfills, as demonstrated by Japan and several municipalities in North America. An integral component of an integrated waste management system is public education. The general people must first be made aware of the information that pertains to them, such as the days that pick-ups are made, where to get recycling containers, what items can be recycled, how to sort them, and so forth. Flyers, newsletters in the local newspaper, segments on local television channels, and phone hotlines can all be used to communicate this information. Instilling a conservation ethic will encourage individuals to take part in the three Rs initiatives, which is an important step. Videos and PowerPoint presentations in schools, posters, buttons, and awards for organizations and corporations that contribute significantly to recycling are all ways to accomplish this goal. The advantages of resource conservation in an environmentally conscious way for participants, their children and grandchildre and the environment must be made abundantly obvious in these programmes. Beginning with the collection of waste from residential areas, apartment complexes, commercial establishments, and industrial facilities, waste management operations begin. This is where the waste management system and waste generators interac These days, the majority of cities run some kind of recycling programme to collecrecyclables alongside ordinary municipal waste. The collection programme is both an expensive endeavor and a challenging logistical challenge. It is crucial that waste producers are aware of what has to be disposed of and how it should be sorted and separate

4.9 WASTE MANAGEMENT SYSTEMS

A modern waste management system has an's important element that is materials recovery facilities (MRF) (Shekdar, 2009). The amount of waste that a municipality can recycle will depend on the effectiveness and adaptability of the MRF. The MRF receives recyclable waste, where the metal, glass, paper, and plastic components are separated, packaged, and kept in storage until they can be delivered. MRFs typically have a drop-off area where people or businesses can bring in pre-sorted recyclables. For materials to be delivered and the finished goods to be picked up, truck access to the building and traffic control are crucial. Even though extra sorting may still be required to eliminate contamination and guarantee that a suitably clean product is given to the customer, it may only be minimally necessary if substantial sorting is done in the trucks that perform the pickup. More complex sorting equipment will be required if the pickup does not perform much sorting. The separation of the various recyclables from the total bulk of material sent to the plant is the most crucial process at an MRF. To separate combinations of materials of various sizes, many types of screens are employed. Vibrating Screens typically constitute flat surfaces on which material is placed and are either horizontal or slightly tilted. Materials of a specific size fall through the screen apertures when the screens are vibrated. Most screen systems enable the use of several screens stacked one on top of the other. Trommel screens are extremely adaptable and consist of a big, cylinder-shaped screen that rotates on a straight axis. Disc Screens are made up of rows of parallel, vertical, revolving discs that lock together. Oversize materials are conveyed along the top while undersized items fall between the discs. Material separation based on density and aerodynamic characteristics is known as density separation. The idea is the same as when winnowing grain.

Cans and other ferrous metals are removed from the waste stream using magnetic separation (Schloemann, 1982). Conveyor belts are frequently used in conjunction with an electromagnet. Aluminum and other nonferrous metals, principally, are separated from plastics and glass using eddy current separation. This approach induces currents in the metal by means of an alternating electromagnetic field. The bits of aluminum are repelled away from the inducing field by the electromagnetic field that is produced by the induced currents, which is in opposition to the inducing field. Even when they use cutting-edge separation technology, MRFs still rely heavily on manual sorting. Due to the fact that manual sorting produces final products of the finest quality i.e., with the least amount of contamination, the current tendency is to combine human and mechanical waste sorting. The final goods created by the various product streams in an MRF are almost always volumetrically reduced in compactors and bundled that can be moved by forklift trucks. An MRF wouldn't be complete without scales. In order to properly account for the various marketable goods and the waste that must be dumped in landfills, they are used to weigh trucks before and after they bring in loads as well as the various streams created by the sorting processes. There is a storage place foe materials at the materials recovery facility. Large storage areas offer for more flexibility and long-term storage in the event that commodity prices are momentarily low. Modern recycling relies heavily on marketing because, without the ability to sell recovered materials or the ability to get a fair price for them, the system would collapse. Ths industry is complicated. It is necessary to locate specialized marketplaces, because commodity values can change significantly over time. In many instances, it might be wise to hire a company that specished compost can be used as landfill cover if no other markets are available.

4.10 COMPOSTING AND PROCESSING

Composting is a specialized form of recycling in which organic wastes are organically broken down under controlled conditions into a product that may be used to benefit the land without harming the environment. Pathogens and other undesirable organisms should all be eliminated throughout the composting process. Composting can improve the texture of light soils, lighten heavy soils, and increase water retention capacity. Since at least the early 1900s, composting has been employed in a systematic way to deal with waste. Composting is a natural process. North America produces a lot of waste and other organic wastes that are easily compostable, composting is a crucial part of a contemporary integrated waste system. Therefore, composting can significantly aid in waste reduction. Most communities can easily implement composting because it is a low-tech, low-cost procedure. Composting is typically appropriate for organic materials, such as non-meat food scraps, leaves, paper, and wood. Plastics and rubber in particular are examples of synthetic organic materials that are rarely biodegradable. The most crucial indicator of the compost's nutrient balance is the carbon/nitrogen ratio. Carbon serves as a vital source of energy for microorganisms, and it is also used to form the architecture of their cells, along with nitrogen. As the composting process continues, the C/N ratio decreases. The final C/N value, which is typically 22:1 influences how the finished compost impacts the soils to which it is applied. More carbon is needed than nitrogen. The microorganisms in the compost will compete with the crops for available nitrogen if C/N is higher than 25:1. The energy source, carbon, is insufficient for the conversion of nitrogen into proteins at compost levels below 20:1.

In this instance, the extra nitrogen is removed by the compost microbes as ammonia, depriving the plants of it and impeding their ability to thrive. By adding either highly nitrogenous materials, such as green plants, or very carbonaceous materials, such as hay and dry leaves, one can adjust the C/N ratio in compost. There are two types of composting programs. The first method makes use of household composters, which are typically 200-liter plastic containers or barrels. These are given to householders, frequently at a reduced cost, and ought to come with guidelines on what and how to compost. Additional assistance can be given through telephone hotlines and volunteer programs where seasoned composters offer assistance and guidance to beginners. Home composting systems are only practical in suburban areas, but they are very successful since waste is diverted at the source and does not need to be picked up or treated by the municipal system. Even if a household composting program is effective, a central composting facility can still contribute significantly to waste reduction. Where household composters are impractical, the central facility can provide services to apartment complexes, businesses, and neighborhoods. It is important to create incentives to encourage landscaping companies, who are major producers of yard waste, to deposit their waste at the collection site. Getting the necessary licenses is a crucial component of constructing a central composting facility, and this includes speaking with any neighborhood organizations that the facility might have an impact on. A sizable amount of land is needed, and this is frequently found at the municipal landfill because there is land there, waste trucks and recycling trucks already go there, there is infrastructure like weigh scales and wood shredders there, and the finished compost can be used as landfill cover if no other markets are available.

Prior to composting, organic waste is processed. Even though composting facilities are a relatively low-tech solution, proper planning and resources are still needed. The most common decomposition techniques are windrow and static pile because they require less initial capital and breakdown takes place aerobically. In aerobic composting, less odor is produced, and temperatures rise to higher levels, killing most pathogens and destroying weeds in the process. One of the earliest and most basic techniques for composting is windrow. Long rows of organic material, measuring 4 to 5 meters wide at the base and 1.8 to 2.1 meters high, make up a typical windrow system.

Actual dimensions can vary and are mostly influenced by the tools used to position and operate the piles. The windrows are turned often, typically once or twice a week, to ensure aerobic conditions and sustain temperatures. A moisture content of 50% to 60% must be preserved. Bulldozers and front-end loaders can be utilized, but turning machines that are more effective and can add water simultaneously have been designed. Anaerobic conditions, which create odor, are prevented by proper aeration. To guarantee pathogen eradication, a temperature of at least 55 °C should be maintained for at least two weeks. The composting process takes around four to five weeks; to ensure that it is thoroughly stabilized, the compost is often cured for a further two to eight weeks.

Numerous organic materials, including yard wasosted using this technique (Francou et al, Linères, Derenne, Villio-Poitrenaud, & Houot, 2008). The materials are arranged in windrow-like long piles. To minimize odor and act as insulation, a layer of screened compost is frequently spread on top of the pile. Either the flooring beneath the pile or the bottom of the pile is covered with a network of perforated pipework. Each pile is supplied with air through the pipe network by blowers, allowing aerobic decomposition to take place. The desired temperature is maintained by regulating airflow rates. To allow for better processing and odor control, the system is entirely or mostly enclosed in contemporary facilities. The process requires more sophisticated equipment than windrow composting, but it doesn't require stirring the material, reduces odors, and offers better process control. The compostable materials are confined in a jar or container. Although vessels come in a variety of forms, there are typically two main types: dynamic and plug flow. The ingredients in the former pass through the vessel without being stirred, whereas the latter stirs or mixes the components while composting. The addition of water and air to the vessels is carefully managed. Because they provide effective process and odor control, quicker composting times, less expensive labor, and the ability to handle food wastes, in-vessel composters are becoming more and more common. They can be placed in urban areas, specifically, to support buildings like hospitals or sizable office buildings. Regardless of the composting technique, the finished product is typically put through a screen to remove large bits and foreign objects and to make sure it looks nice enough to be sold. Leachate may be created in large amounts because composting requires maintaining high moisture levels. A leacte collection system, a water treatment facility, and an environmental monitoring programme may be required, depending on the kinds of materials being composted and the permitting requirements.

A crucial component of the plant is a storage, packaging, and loading space because the finished compost must be sold to a market. Landfills are often only utilized as a last resort for the disposal of sludge when no other viable options, such as fertilizer for agricultural areas or fuel for incinerators, are available.

CHAPTER 5

Electronic Waste Management

CONTENTS

5.1 INTRODUCTION

Concern over the detrimental effects that industry and its products are having on society and the environment in which we live has grown over the past several years. As a result, the idea of sustainability and the desire to act more sustainably have drawn more attention. The world's population is expanding quickly, and wealth is typically increasing, which has led to an unsustainable increase in material, energy, and other resource use. With the growing recognition that human activity has a substantial impact on issues like global warming, there is an obvious need to change how society consumes and frequently wastes precious resources. In other words, we need to act more sustainably. The European Commission (EC) has recently introduced a package of Producer Responsibility legislation to address the issing recognizing that society as a whole and the electrical/electronics industry in particular need to act in a more responsible and sustainable manner (Mrkajić etaro, 2018). The European Commission is driving this in order to fulfill a number of goals targeted at a more sustainable approach to resource usage and a decrease in the amount of waste going to landfills. Along with outlawing the use of some hazardous materials and lowering energy usage throughout the product lifespan, it also tries to divert end-of-life products for reuse, recycling, and other types of recovery. It is obvious that the electronics sector needs to operate more sustainably in order to both meet the demands of the ever-tougher legislation and those of the customers who want that industry to adhere to high environmental standards. The adoption of innovative production techniques, the use of novel materials, and the creation of improved recovery and re-use strategies at the end of life are all ways that the electronics sector might accomplish these goals. Even if the industry can often already accomplish this, there are some longer-term prospects that can only be met through additional research and development.

5.2 WASTE ELECTRICAL AND ELECTRONIC EQNT WEEE

Nearly 2 million tonnes of Waste Electrical and Electronic Equipment (WEEE) are produced annually in the UK. Sales data from 2003 used as the beginning point for data compilation in past research on WEEE, expressed as weight and units for the categories outlined by the WEEE Directive. Manufacturers, merchants, trade associations, and market research organizations provided information. According to the findings, 93 million pieces of domestic equiment totalling 939,000 tonnes were dumped in the UK in 2003. Without a doubt, WEEE poses a severe issue, not only in terms of how its final management of treatment and disposal s handled, but also in the context of sustainability and the waste of priceless and limited resources. The main goal of the 'Restriction of Hazardous Substances' or RoHSimain goal is to preserve human health and the environment by limiting the use of some hazardous substances. These substances are specifically polybrominated biphenyls, polybrominated diphenyl ethers, lead, mercury, cadmium, hexavalent chromium, and cadmium. In the UK, RoHS was enacted into law in August 2005, and the prohibited use of the listed hazardous compounds began in July 2006. Manufacturers, retailers, distributors, and recyclers of electrical and electronic equipment have all felt the effects of the RoHS Directive and continue to do so. Producers are responsible for making sure that the goods they place on the European market are free of prohibited materials and that they adhere to the Directive's guidelines. A producer may be required to remove products from the European market if it is discovered that they include certain prohibited elements. All of the product categories listed in the WEEE Directive are covered by the RoHS Directive, with the exception of the medical and monitoring and control categories. The RoHS Directive stipulates that a material shall not be present in a homogeneous material over a certain weight (%) because it is impossible to remove every single atom of a substance (Johansson & Luttropp, 2009). Except for cadmium, where the level is ten times lower at 0.01 percent, this amount is established at 0.1 percent by weight for each of the compounds that are prohibited. The RoHS Directive has promoted the introduction of analogous laws all over the world, even though it only applies to products sold in European member states. The so-called China RoHS, which forbids the use of the same materials as the European RoHS Directive but implements the rules in a completely different manner, is perhaps the next most well-known piece of legislation of this type. A variant of RoHS known as super-RoHS since it covers 18 different chemicals rather than just the six listed in the European RoHS Directive has recently been announced as something Norway is contemplating implementing.

5.3 PROCESSING

Manufacturers will be obliged to evaluate the energy and material consumption, pollution, emissions to the air and water, projected waste, and recycling/reuse for each step of this process. The EuP Directive (EuP), which places focus on all facets of a product's lifetime, from eco-design to end of life, urges the electronics sector to adopt a more holistic approach to how it develops its goods. In order to improve a product's environmental performance and achieve more sustainable product development, eco-design principles should be encouraged. This will result in the incorporation of environmental issues during the product's design process. Better developed, more effective products should be able to provide manufacturers and consumers with advantages on both an economic and resource-saving level. A final draft of the EuP Directive 2005/32/EC was approved by the European Parliament and the Council in July 2005. As part of the so-called "fast-track comitology procedure," actual measures are being chosen on a product-by-product basis under the guidance of a designated panel of experts from EU member states. Heating, electric motors, lighting, and home appliances are among the top products. This directive will eventually apply to all energy-consuming products, excluding motor vehicles, which are estimated to be responsible for 40% of the carbon dioxide emissions that cause global warming and are to be reduced under the Kyoto Protocol. Undoubtedly, the use of new materials and procedures will become more and more crucial in the effort to comply with the law.

The new European system for regulating chemical safety is known as REACH (Registration, Evaluation, Authorization, and Restriction of Chemicals) and went into effect on June 1st, 2007. They could have a substantial influence on end users as well as chemical producers farther down the supply chains of the sector. The majority of enterprises, including many that would not typically consider themselves connected with chemicals, would be impacted by REACH. It is anticipated that there will be effects on electrical and electronic items, and it's possible that other materials won't be allowed in addition to those listed in the RoHS Directive. Despite the fact that the list of these compounds is not anticipated to be released until 2009, there are signs that some plasticizers and flame retardants used in electrical equipment may be banned. The capacity to recycle materials from products marketed before the REACH Regulations went into effect may therefore be impacted by the prohibition of such materialdfester, & Wurm, 2019). Integrating cost-effective materials compliance procedures across multi-tiered global supply chains will be a major problem for many large multinational corporations. Given that global supply chains frequently originate in the Far East, there will undoubtedly be a growing need to reduce the amount of supply chain confusion that could occur from the implementation of inconsistent regulations in various states, nations, and geographical areas of the world. There are already problems with certain egions and their' access to RoHS-compliant components but not others. Since one obvious benefit would be a decrease in the requirement to produce region-specific products, the business should ideally work toward better levels of agreement with regard to environmental legislation. It's crucial that any new environmental legislation, no matter where it's enacted, doesn't have a significantly different scope from other similar legislation in a different area. As more environmental legislation is put into effect over the next few years, achieving some degree of harmonization through an international standardization process will become more crucial.

5.4 POLICY MAKING

It will be necessary for industry and policy makers to collaborate more closely in order to achieve a unified approach to environmental regulation, particularly legislation governing aterial restrictions, so that legislators are aware of the environmental trade-offs associated with material subtituton. Industriesy needs to take a more active role in negotiating how the costs of resource efficiency and sustainability best practices can be absorbed into the economy without endangering inflation and jobs. In order to demonstrate compliance, as well as to know the environmental impacts of materials and potential trade-offs of alternatives, it will also be necessary to create and standardize good scientific methodology and testing processes. Designers will need a thorough and standardized framework system to help them determine the status of a given substance with regard to its environmental classifications in categories like regulatory restrictions, EU risk phases, and persistent organic pollutants, as well as to help in material selection in the interim or in the absence of such a harmonized approach. Even if there have been some improvements in recent years regarding WEEE treatment choices, there is still a need for innovative technologies that can improve WEEE treatment effectiveness in a moremated manner with less fewer manual interventions. A certain amount of manual separation may often be used in conventional treatment methods before a range of more automated comminution and separation processes. There was a lack of clarification regarding the precise elements of the WEEE Directive's treatment requirements in the period leading up to its implementation. For instance, it wasn't clear exactly when printed circuit boards (PCBs) and liquid crystal displays (LCDs) needed to be removed from the waste stream throughout the recycling process. Although there are ten different kinds of WEEE listed in the WEEE Directive, the actual number of segregated streams that are actually collected means that different product types are not separated as effectively as would be ideal for later optimized treatment.

The material composition of WEEE is one of the primary variables that influences the selection of the most appropriate technology for recycling. Even within certain product categories, there are distinctions in the sorts of equipment that fall under each of the ten WEEE categories. A very appropriate example is the quickly accelerating switch from CRT-based televisions and displays to those using LCDs. Laws like the RoHS Directive are causing changes in many other areas as well, including the materials that are used in common in many electrical and electronic equipment. The changeover from lead-based to lead-free solder, which has already happened and has been required for many items since July 2006, is the most well-known example. While this won't have an effect for many years on some long-lasting products, like televisions, the introduction of short-lasting products, like mobile phones, will cause the composition of printed circuit board (PCB)-related scrap to change and likely expose a wider range of metals than the tin, lead, and copper typically found in traditional circuit boards (Wang & Gaustad, 2012). The ban on some brominated flame retardants, cadmium, mercury, and hexavalent chromium will also cause compositional changes that will signal the entry of a greater variety of elements into the waste stream. This in turn will have implications for any recycling technology created to address specific waste streams for each of the ten categories covered by the WEEE Directive. In contrast, it will also become more crucial to separate RoHS-compliant products from non-compliant ones because, in cases where materials are being recovered and recycled for use in new electrical and electronic products, the presence of prohibited materials will make the treatment and recovery processes more difficult and expensive, which could lead to contamination problems.

It has been known for some time that there is a need to reduce the types of plastics used in electrical and electronic devices in order to permit more effective recycling. The majority of types of EEE contain various volumes and types of plastics. Compatibility problems exist not just between different classes of polymers but also between the numerous products made for each class, which can make the situation even more difficult. It should be noted, however, that it is rather typical to see many more types of materials being utilized in specialized applications, and electronics that incorporate circuit boards will also contain a variety of thermoset materials, like flame-retarded glass-reinforced epoxies. The kind of polymer, the cost in comparison to raw material, and the amount of labor necessary to create a new material with the needed purity and quality all have a significant role in how useful recycled plastics can be. For instance, the cost of recycled materials may increase dramatically if the components must be separated and any impurities like labels, screws, and fasteners must be removed. The effects of recycling polymers with brominated flame retardants must also be taken into account. Due to their enduring presence in the environment and potential hazards to human health and safety, these are losing popularity. It's interesting to note that recycled plastics in Japan have been used as conduits for wires that run alongside railway tracks. The wide variety of plastic materials on the market and their inherent incompatibilities highlight the need for designers to use fewer different types of materials when creating new items. Manufacturers have occasionally employed alternate materials, such as aluminum, in some situations, such as laptop computer cases, to help with easy recycling after use.

The amount of information that is now available about the precise material composition of EEE is rather small. It is obvious that the method of product segregation must be defined, yet best practices vary and are continually developing. However, it will be critical to characterize each waste stream in terms of its product content and individual material composition in order to successfully adopt appropriate, category-specific, focused recycling methods. Regarding the supply of raw materials to be processed, each recycling technology or process will operate at its most effective level. It would be easier for people who gather and aggregate particular groups of items to control the composition in order to achieve increased efficiency if they have feedback regarding the material composition necessary to achieve maximum efficiency. The viability of the many procedures that can be used to address WEEE is significantly influenced by socioeconomic conditions (Pagdee, Kim, & Daugherty, 2006). They also have a big impact on the approaches that can be used now for material recovery as well as the ones that are most likely to be successful in the future. For instance, recycled polymers from end-of-life electronics are currently hardly ever used in any significant capacity, despite the fact that businesses like Axion Recycling Ltd., which recently put into operation a facility in Salford, have developed procedures that are starting to make it possible to provide high-quality recyclables. In contrast, the big electronics businesses play a crucial role in Japan's more integrated approach to electronics recycling. This indicates that there are internal projections for the usage of such recycles within vertical supply chains. There is technology for making higher-quality polymers, but it is hindered by collection processes that frequently result in contamination of the basic granulated material with other materials such as metals, paper, dirt, and other plastics. This is partially a result of how business is conducted in Europe, where there is little interaction between producers and recyclers. Because the big electronics manufacturers in Japan really own the recycling facilities and are eager to get high-quality recycled materials for use in their own goods, this is the outcome of a mix of historical and cultural reasons. Internal supply chains don't really exist in Europe, and especially not in the UK, unlike in Japan. Therefore, establishing supply chains both before and after recycling is a crucial requirement for WEEE recycling facilities in the UK in order to optimize the processes f material recovery, recycling, and reusing.

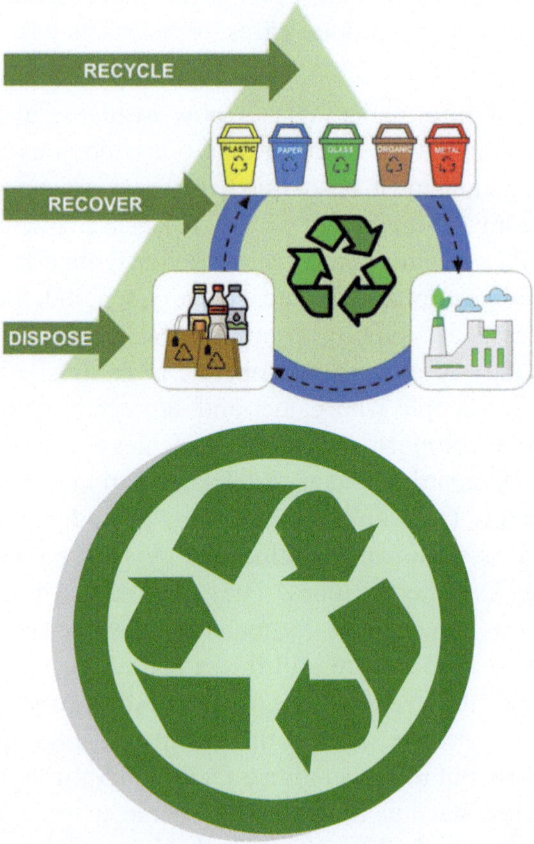

Figure 5.1. Recycling. (Source: Wiśniewska P, Hejna A, Saeb MR. Recycling and Processing of Waste Materials. *Materials*. 2023ikipedia/commons/8/8b/Symbol_recycling_vote.svg, Creative Commons License).

5.5 RECYCLING

There will also need to be significant changes in the economics of recycling (Figure 5.1) if a business is to adopt the new recycling technologies that will unavoidably be required as regulation forces the recycling of more and more electronics waste. The exports of WEEE from Europe in recent years, which have sparked controversy and embarrassed some well-known electronics manufacturers, are evidence of this. The amount of recycling activity evidently depends on other global conditions as well. For instance, several of the metals included in WEEE have seen huge increases in value recently, partly as a result of rising demand from developing nations like China. Since the turn of the century, the price of copper has tripled, creating an opportunity for the introduction of newer technologies that will, ideally, also provide more high-quality recyclate and allow for more recycling to be done in the UK. One significant element impacting decisions about the expansion and investment in metals recycling, which is presently a critical component of the total WEEE recycling process, will be whether these gains in the value of basic materials will be sustained. Oil and energy prices also need to be taken into account, as they have an effect on how raw materials used in electronics are produced. They have an impact on how much it costs to operate recycling facilities as well as to transport WEEE to the recyclers. Of course, it is ideal for WEEE to travel as little as possible from the point of disposal to the point where it is recycled and treated before being used by another party. A WEEE recycling facility should process as much locally sourced WEEE as possible and provide the recycled material to nearby end users from a sustainability standpoint(Jang, 2010)generation of waste electrical and electronic equipment (WEEE. This could not always be achievable, especially given the process's lack of integration as noted above and any potential local uses for the recycled material. However, utilizing the right lifecycle techniques, all of these potential recycling paths need to be taken into account more comprehensively. When the materials from end-of-life electronics can be used to create new products in the Far East, for instance, shipping them there for recycling and material recovery may make sense in some circumstances.

The need for and capacity to implement new technologies in electronics recycling are both influenced by a wide range of socioeconomic factors. Due to their interconnectedness and mutual dependence, there are several complicated interactions and factors that affect what happens to WEEE. Comparisons with what is accomplished in nations like Sweden and Japan ampl highlight the need to achieve greater levels of re-use, recycling, and recovery from UK WEEE. It is already required by the WEEE Directive that particular targets be fulfilled for each category of WEEE, and it is anticipated that these may get stricter in the future as a result of legislation like the Energy-using Products Directive, among other things. Socio-economically, there is much that could be done in the UK to support the adoption of these technologies and, consequently, the better processing of WEEE materials. Companies now pay for the recovery of their products at the end of their useful lives as a result of producer responsibility law and the producer pays principle. By establishing mandated recycling and recovery targets that must be funded by the manufacturers, these legislative initiatives aim to decrease the amount of waste that is dumped in landfills. This new approach to waste management legislation within the European Union has been aimed at packaging waste, end-of-life vehicles, and, more recently, electrical and electronic products. The necessity to create new, economical, and effective logistical methods to facilitate transportation of WEEE from collection sites to the recycling facilities is one of the key challenges in fulfilling these te-back criteria.

Different WEEE collection programmes have been formed around Europe as a result of the freedom granted to European Member States to specify the general needs of their unique implementing legislation. This is in stark contrast to the strategy used in Japan and has surely raised the unneeded complexity and cost for the companie Producers, trade associations, compliance programmes, and local governments attended meetings with government representatives during the WEEE Directive implementation phase to discuss the best ways to allocate and transport WEEE to recyclers in the UK. As a result, and since the Directive's introduction, the UK's methods for collecting and transporting WEEE from collection sites to treatment facilities have grown and changed. However, representatives from the waste management, transportation, and producer sectors have made it clear that the UK's delay in enacting the WEEE Directive into law created significant problems for logistics planning and made it very challenging to make the proper investment in data collection and capital equipment, among other things. Given that many homeowners leave large, bulky appliances at CA sites, the decision to define Civic Amenity (CA) sites as collection points and delegate the subsequent collection and recycling of WEEE to the producers was one of the more easily understood ones relating to WEEE allocation and collection. There is a great amount of room to increase WEEE collections at CA locations based just on available space, and such a method would significantly aid in the Directive's implementation. More than 75 percent of the WEEE that is recycled comes from large, high-metal domestic appliances i.e., washing machines, refrigerators, freezers, etc. that make up the majority of the infrastructure for WEEE separation at CA facilities. This high recycling rate may be partially explained by the high value of the metals recovered in comparison to the lower reprocessing costs, which are further reduced by the fact that the facilities used to shred end-of-life goods are occasionally also used to recycle end-of-life vehicles. Due to the specialized handling required by the Ozone Depleting Substances Regulations, refrigeration equipment is one of the additional appliances that is separated at CA sites. Additionally, it should be hhlighted that there are a few smaller-scale programmes for sorting small appliances to promote re-use schemes.

The WEEE Directive mandates not oy the establishment of efective collection programmes that achieve its goals, but also that WEEE be collected and transported in a way that encourages the reuse and recycling of parts and materials from appliances that can be recycled. However, packing as much product onto a vehicle as you can is necessary to achieve cost-effective WEEE transportation. This typically entails the use of powerful machinery, such as JCBs, which can seriously harm the equipment. The Japanese approach was different in that WEEE was not seen as waste but rather as a rich source of raw materials. In order to minimize damage during transit, the WEEE was streamed by item type at the collection station and carried in specialized cages on specifically constructeraulic vehicles. This process resulted in higherthan -than-anticipated recycling rates and increased yields of higher-quality materials.

There doesn't seem to have been much research done on the types and quantities of WEEE produced in the UK. Organizations engaged in the production of electrical and electronic products operate and own the schemes. Denmark is an exception, as municipal anregional authorities run the WE collecting programme. Initiating collection programmes in 2005 were Austria and Ireland, and during the past few years, other EU member states have incorporated the WEEE and RoHS Dnto natnal law and registered coection collecting programmes. The collection programme in Germany began in the first quarter of 2006. So nations, likeun multiple programmes under which various programmes collect various WEEE categories. Japan is well renowned for its aggressive effort to recycle electronics, although its strategy differs significanring the end of their useful lives is highly helpful.

5.6 LAWS

The Home Appliance Recycling Law (HARL), which went into effect in April 2001, ended up serving as the primary impetus for the substantial level of recycling activity that has been attained in recent years (Ongondo, Williams, & Cherrett, 2011)regulate and prevent WEEE are comprehensively examined. The findings indicate that for (i. Televisions, refrigerators, washing machines, and air conditioners are the four main categories of home appliances that the HARL addresses. As a result, it differs from and is less complicated than the various national programs that are being planned and put into place in the UK and other European nations in response to the WEEE Directive.

It's interesting to note that the procedure is effective because a lot of appliances are recycled and a lot of the recycled components are put to good use. The fact that individual consumers are prepared to pay the appropriate recycling cost when an appliance is sent for recycling is further evidenced by this. Given that WEEE is currently routinely dumped outside of permitted recycling facilities and Civic Amenity sites, it is impossible to imagine that this strategy would be effective in the UK. The implementation of such a consumer fee-based plan in the UK is likely to result in an increase in the unlawful dumping of even more WEEE given the stark cultural differences between the UK and Japan. There are 46 home electric appliance recycling facilities operating in Japan that handle and process the HARL-covered end-of-life appliances. The adoption of a ticketing or voucher system that guarantees the traceability of an end-of-life appliance on its route from the user to the recycler is a crucial component in the successful implementation of the Home Appliance Recycling Law. It has been discovered that using the ticketing system to monitor and track appliances nearing the end of their useful lives is highly helpful.

The Home Appliance Recycling Law imposes an "old for new" requirement on Japanese shops, similar to the take-back rules for WEEE in Norway, Sweden, the Netherlands, and Belgium. This implies that one item must be taken back from the customer for each item that is sold. This can be a previously sold item that is identical to it or another product. In addition, the law allows manufacturers to hire third parties, such as the Association for Electric Home Appliances (AEHA), to perform collecting tasks on their behalf. The local government or the AEHA provides collections in rural areas without big appliance shops. According to the Japanese Ministry of the Environment, retail establishments receive 80% of the recovered appliances that are collected. Retailers, local governments, and other approved organizations are required to transfer the collected appliances to two manufacturer-operated consolidation centers after collection. The Japanese industry has established these two consortia because the HARL says that manufacturers have individual accountability for their products (Goldman & Nagel, 1993). Each consortium's manufacturers must also set up regional consolidation centers and arrange for the transportation of collected goods from these centers to recycling facilities. The consumer is responsible for paying a fee when an appliance is consigned for recycling. This is a totally different strategy from the upfront European funding model, in which recycling costs are either built into the price of the product and invisible to consumers, or they are disclosed in advance as recycling fees that are noted on the product receipt at the time of purchase. When taking their used goods for recycling in Japan, consumers must pay a collection charge that is determined by the merchant or other collection provider. Additionally, they must purchase a recycling ticket, which is available from post offices and some retail establishments.

The manufacturers receive a monthly transfer of the collected recycling fees. They are meant to cover the expenses related to running the regional consolidation centers, shipping goods to recycling facilities, and carrying out the recycling process itself. The makers are liable for paying the leftover expenses because the fees do not fully cover the costs of recycling. The tickets include information about the appliances, the retailer's name and the manufacturer's name. With the help of the information and reference number on the ticket, customers can check the status of their appliance, such as whether it has been sent back to the manufacturer or is currently being recycled. Customers may fully trace their appliance using this rather simple approach. They can use a checking system that keeps track of the status of all collected electrical appliances to enter the AEHA website and determine the status of any electrical appliance that has been submitted for recycling. Consumers must agree to cooperate with retailers in properly transferring used appliances to assure recycling as well as to pay the necessary fees for the transfer and recycling of those products. Similar to this, when selling the same type of home appliances to customers as well as when selling used goods that they themselves previously sold to customers, shops are required to accept them back. They must also transmit them to the appropriate importers or manufacturers. The Japanese government has established recycling standards, and manufacturers and importers are required to accept appliances for recycling at approved take-back locations and to recycle them in accordance with those requirements. The Japanese model only deals with four specific forms of WEEE, which is one of the main distinctions between the European and Japanese approaches to WEEE recycling. In contrast, the WEEE Directive in Europe covers ten product categories and aims to include almost all electrical and electronic waste. The Japanese government believes that the four product categories covered by the law account for 80% by weight of all discarded electrical and electronic equipment, despite the fact that they have focused on a far smaller variety of products than in Europe. Although it should be noted that the Japanese were among the first to adopt lead-free manufacturing for their electronics, way ahead of any legislation in Europe, it is interesting to note that there is no equivalent complementary Restriction of Hazardous Substances Dctive-type legislation in Japan. The distribution of WEEE in with respect to the location of the manufacturing plant is one of the major obstacles to manufacturers recycling WEEE. Many WEEE producers are located far away from both their markets and the waste they produce. They find it challenging and expensive to run take-back locations exclusively for their appliances as a result. Retailers and distributors are ideally positioned to collect WEEE since customers can bring small appliances to their local store when purchasing or picking up new appliances, or they can have old appliances collected while delivering new ones. The expense of collection sites and storage facilities, which in addition to having limited capacity, would also have problems for security and health and safety, would need to be covered by shops.

Major WEEE collectors in Scotland are the local governments (Darby & Obara, 2005)with estimates of up to 20kg per person per annum. A wide variety of WEEE is discarded by consumers, often in different ways depending on size with small items (e.g. toasters. The fact that the appliances are composed of various materials, making disassembly labor-intensive and consequently expensive, is the fundamental issue that scrap metal merchants and material recyclers face. A low-quality end product results from the resultant metal being frequently contaminated with plastic and other insulating materials, etc. This indicates that there isn't a ready market for such products, therefore processing these equipment frequently isn't financially viable. In the past, any discussion about WEEE recycling would have typically focused solely on the recovery of metals. This has been particularly true for particular electronics product categories that include significant amounts of precious metals like gold and palladium as well as other metals like copper, tin, and silver. Even with bigger, more popular electrical devices like washing machines, recycling has traditionally meant recovering copper and steel. However, there is a distinctly defined hierarchy that may be used when processing WEEE in terms of both more sustainable methods and recovered value. Therefore, the main potential for reusing specific functional units, modules, and components is inside the broader WEEE re-use, recovery, and recycling hierarchy. Nowadays, there are a lot of businesses that really restore electronic items for reuse; this is especially true with mobile phones and personal computers. In order to provide low-cost access to items for those who would not otherwise be able to buy new products, these organizations are frequently run as social businesses. This has made it possible for more individuals to access computers and the internet, including those in the UK and, forbile phone has one of the biggest effects on the environment.

5.7 ELECTRONIC PRODUCTS

Repurposing specific types of electronics is an intriguing and relatively new idea where it is not viable to fix and reuse electronic equipment. Although this method has not yet been widely applied, it may be useful in other contexts due to the rising sophistication of some electronic devices, such as mobile phones. The most recent mobile phones have extremely sophisticated processing and communications capabilities that may be used for novel applications; however, they are frequently discarded after only a few months of use, usually between 12 and 18 months. Given that, there were more than 2 billion mobile phone users worldwide in 2005, this amounts to a significant loss of energy, raw materials, valuable components, and electrical functioning. According to lifecycle assessment (LCA) research, the energy consumed to produce the components of a mobile phone has one of the biggest effects on the environment.

Recently, research has been done to determine whether it is feasible to turn mobile phones into general smart processing units that may be used by companies that make new electronic equipment. In order to use redundant units as main components in a variety of applications, such as those requiring smart monitoring, sensing, and telemetry capabilities, they must first be converted into general-purpose modules. Applications might include enhanced mobile medical diagnostics for usage in rural areas as well as alarms and monitoring. Even while this method only works with a select group of complex, high-volume electronic items, it does present some intriguing alternatives to the usual disposal and recycling methods, and it is surely worth further study. Before sending the rest of an electronic assembly to more traditional recycling, it could be possible to recover some working electronic components where refurbishing and repurposing are not appropriate. This makes it possible to extract more value from things like the assembled printed circuit boards that are included in most goods, for instance. For the automatic deconstruction of components from trash/waste, unnecessary, or broken PCB assemblies, the Austrian organization SAT created a methodology. The core of SAT's technology was a laser desoldering unit connected with a robotic vacuum arm for the removal of specific components, as well as an automated optical recognition scanning system that identified components acceptable for recovery. The recovered parts could be utilized again after removal and testing. The NEC Group in Japan has also worked on creating a mechanically-based automated disassembly technique.

The above-mentioned technologies, while highly intriguing, are most likely only going to be able to handle a small percentage of the entire volume of electronic waste. These pathways won't be practical for a big portion of the vast number of electronic waste products since many forms of WEEE don't actually contain the kinds of materials that are cost-effective to recover and reuse. The numerous low-cost consumer goods, including kettles, toasters, radios, CD players, and similar items, serve as typical examples. These don't have any large electronic components that are valuable, thus they will probably be processed using methods that don't considerably deviate from those that have been used for a long time. In essence, these require using a variety of sorting processes to distinguish between the metallic and plastic fractions. Iron and nonferrous metal fractions can be separated using more advanced procedures, but in the end, the metals will go to a refiner that will separate the constituents to create high-purity recycled metals. The plastic fractions can also be partially divided into different types, however, in many circumstances, these materials may also wind up being sent to the furnace as part of the metal refining process. Therefore, rather than just sending these materials to a refiner, a crucial question that needs to be answered is whether or not they can be better separated and used in alternative and creative applications. Much more valuable than materials that have been mixed with other materials and may be both polluted and dirty are materials that can be kept clean and uncontaminated. This makes it possible to manually disassemble and separate recyclable pieces. For instance, it can be directly recombined and utilized again or combined with raw materials to create a variety of products with qualities that are almost exactly the same as those of new material. What to do with plastics from end-of-life

EEE that include brominated flame retardants is one of the major problems. The Japanese have cable networking for signaling wires along the side of the Japan's railways using recycled flame retardant plastics from WEEE (Dimmen, Steindal, & Basu, 2023). These huge moldings utilize a lot of recycled material, provide an alternative to conventional concrete products, and produce a stable material with a long lifespan. Another advantage is having a material that is flame-resistant. The presence of halogens, heavy metals like cadmium and lead, and brominated flame retardants can then be determined by the subsequent changes in the color of the spark. For polymer recyclers, this kind of hand-held device can be a very helpful tool.

Axion Polymers, based in Salford, has produced a new RoHS-compliant grade of high-impact polystyrene that is fully made from used TV and computer monitor plastic, providing a recent example of the effective recycling of polymers from end-of-life electronic products. Axion's proprietary procedure separates the portion of discarded casings that contains brominated flame retardants to create a purer substance called Axpoly PS02. Also noteworthy is the recent announcement by MBA Polymers UK of plans to construct a 60,000 tonne capacity plant for the recovery of polymers from enhanced shredder residue. When vehicles, electronics are recycled, shredder residue is created, which is a complicated mixture of materials rich in polymers. According to estimates, the UK creates about 3 million tonnes of plastic annually from the disposal of end-of-life devices like automobiles, gadgets, and electrical appliances. Because the recovery procedures are seen as being too complex or expensive to operate, the plastics from them are often landfilled or burned at high economic and environmental costs. Between two and three tonnes of the greenhouse gas CO_2 are kept out of the atmosphere for every tonne of raw plastic that is replaced. Future interest is increasing in the use of polymers derived from non-fossil fuels in electronics applications, such as mobile phone casings. For instance, interest in polylactic acid (PLA), a bioplastic made from corn, has been developing recently as a potential substitute for traditional plastics made from petroleum. PLA had not previously been utilized in electronic equipment due to its inadequate heat resistance and strength.

The method was largely focused on the separation and recovery of metals for refinement, but it also produced a polymer-rich fraction that was recycled and utilized to create a metal-free epoxy glass-fiber composite powder. The disposal of CRT-based televisions and monitors is a significant source of trouble for electronics recyclers (Andreola, Barbieri, Corradi, & Lancellotti, 2007)in the last decades it has been noticed an appreciable increase of this residue, consisting of about 80% of television sets and computers containing end of life (EOL. Since traditional CRT manufacturing has been replaced by LCD technologies, one of the primary uses for recycled glass containing lead has fallen off quickly. In the ceramics sector, it has also been demonstrated that glass from used CRTs may be used as a raw material in goods that are fired at low and high temperatures. CRT glass that has reached the end of its useful life has also been used to concrete mix. This application encapsulates the lead-containing CRT glass by substituting it for sand when creating a concrete composite. Glass, biopolymers like gum, and a cross-linking agent can be used to create concrete composites with compressive strengths of up to 6000 psi. Several items created from discarded or outdated circuit boards have been commercially released. The most important examples of these are clocks, coasters, briefcases with aluminum frames and circuit board panels, and ring binders with circuit board sides.

A more informed approach to end-of-life product segregation, which may have to start as early as the design stage, is essential to maximizing the use of recycled materials. This entails not only sorting things and keeping them in good condition before disassembling and recycling them, but also developing products that are simpler to handle at the end of their useful lives and have fewer types of materials and fasteners. The least-expensive approach of simply shredding mixed WEEE items in bulk, as was done in the past throughout Europe, inhibits opportunities from being developed to maximize recycling rates and, as a result, to promote additional recycling of valuable materials. Electrical anc products are made of a variety of materials, some of which are recognised recognized to pose risks to the health of personnel involved in providing end-of-life care. End-of-life electronics have a number of potential hazards from a health and safety perspective, including possible risks to the environment and workers in the recycling business. Examples include failing to employ the necessary personal protective equipment and exposing recyclers to hazardous contaminants by using ineffective treatment methods. Environmental contamination can also result from improper handling and disposal.

5.8 TELEVISION WASTES

There has been an increasing trend to switch from older CRT-based televisions (Figure 5.2) and computer displays to more modern models that use liquid crystal and plasma display technology.

Figure 5.2. CRT-based televisions. (Source: https://upload.wikimedia.org/wikipedia/commons/0/00/Tunisia-3433_-_Want_to_buy_a_TV......_%287847360164%29.jpg, Creative Commons License).

An estimated 2 million CRT-based televisions and extra CRT-based monitors are disposed of annually in the UK, producing roughly 100,000 tonnes of CRT glass annually. As a result, there are many CRTs that need to be treated and that also contain a number of dangerous elements. Several potentially cancer-causing substances, including lead, barium, phosphors, and other heavy metals like cadmium and mercury, are found in CRTs. When handled correctly in a controlled environment, these do not present any significant health or environmental risks, but the dismantling, separating, and disposal of CRTs in an uncontrolled setting without the required safety measures can present significant risks for workers engaged in this type of process and result in the release of toxic materials into the environment, including soil, air, and water. Japanese recyclers, on the other hand, are renowned for using a much safer and more controlled method. There are many different recycling techniques used to deal with end-of-life CRTs, ranging from those that obviously lack the controls and processes required to prevent worker exposure and environmental contamination. Since a traditional CRT-based television has a lifespan of around 14 years, there will be plenty of CRTs available for recycling for many years to come (Mudali,

Patil, Saravanabhavan, & Saraswat, 2021). The end-of-life materials discovered with LCD and plasma-based displays will be different from those discovered with traditional CRT-based displays. Several variables will determine whether or not the situation will improve or deteriorate from the perspective of health, safety, and the environment. For instance, there is a trend toward wide TVs and displays, which could require treating larger amounts of materials. Tneral view appears to be that these more modern display types will provide less fewer issues than classic CRT displays. Less heavy metals will also be present in LCD screens, and they won't require the use of phosphors like those used in CRT displays. It should be remembered, nevertheless, that many LCD screens employ backlights that contain mercury that must be handled carefully. Mercury backlights can today be treated in a variety of ways, but it can often be challenging to remove them from the back of a display without them breaking and exposing employees to both mercury and the phosphors that produce the light in the tube. Some large-area LCD panels have more than ten mercury backlights, so designers must think about how to remove them more easily at the end of their useful lives. It's interesting to note that research has generally demonstrated that the liquid-crystal compounds employed in these displays have very minimal toxicity. It has been recommended that there are no particular standards needed for the disposal of LCD screens based on the findings of experiments to assess the eco-toxicology of liquid-crystal materials. Therefore, LCDs may have less of an impact on human health and the environment than the CRTs they replace if the problem of the mercury-containing backlight can be solved by replacing them with newer technology. It's also important to keep in mind that alternative light sources are increasingly taking the place of mercury backlights in various applications, raising the possibility that they won't be needed in the future.

Treatment of end-of-life electronics containing brominated flame retardants (BFRs) is another area that may raise both health and environmental concerns. BFRs clearly serve a very valuable purpose and prevent many lives from being lost in fires, but in recent years, they have also drawn increasing attention due to their persistence in the environment and their effects on human health. Thermosetting polymer systems, like the epoxides used in circuit board laminates, and non-reactive polybrominated diphenyl ethers (PBDEs), which are utilized in thermoplastic polymers, are both types of BFRs that can be found in electronics applications. According to research, PBDE levels in people double every two to five years and are substantially greater in North America than on other continents. While

PBDEs are known to migrate out of polymers, especially when they are heated, reactive BFRs, including those based on tetrabromobisphenol A, have relatively little volatility. Therefore, any recycling procedure that involves heat may result in PBDE emissions, which could have an effect on the environment as well as recycling personnel(Hahladakis, Velis, Weber, Iacovidou, & Purnell, 2018). It's interesting to note that there is evidence to suggest that restrictions on the use of brominated flame retardants can raise the possibility of environmental damage from fires involving electrical devices that are no longer, or are flame retarded less efficiently.

The problem of brominated flame retardants needs to be addressed because there is an increase in interest in recycling and reusing polymers from end-of-life devices. For instance, since some procedures expose the polymer to high temperatures that might cause the emission of BFRs and their degradation products, it is necessary to put in place stringent controls for personnel involved in the recycling of polymers containing these compounds. The industry has taken steps to create a coordinated best practice strategy in response to the mounting concerns around the usage of brominated flame retardants. The Voluntary Emissions Control Action Programme (VECAP), which was developed by the bromidant business, serves as a good illustration of this strategy (Chen et al, Li, Liu, & Zhao, 2012). Through collaboration with small and medium-sized businesses, VECAP was established to control, monitor, and reduce industrial emissions of BFRs into the environment. Effectively, it aims to establish best practices before they become necessary under the "registration, evaluation and authorization of chemicals" or REACH requirements. Manufacturers and consumers have been collaborating to develop and disseminate best handling practices for deca-BDE in order to reduce environmental emissions up to this point. Deca-BDE levs in the environment are said to have decreased following the implementation of this programme. While minimizing the negative effects of end-of-life electronics on the environment and human health is one of the main goals of the WEEE and RoHS Directives, there are still issues with WEEE export to developing nations like China and Africa. Although it is technically against the law, this practice continues to take place, and large amounts of unwanted electronics are processed in these nations under unfavorable circumstances, exposing recyclers and the environment to harmful elements.

5.9 MERCURY

The amount of mercury used in electrical and electronic equipment is thought to be 22% of the total. It is used in batteries, medical equipment, mobile phones, fluorescent lighting, sensors, relays, switches, and thermostats. The use of mercury has increased in flat-panel display backlights as liquid crystal displays replace traditional cathode ray tubes more frequently. Therefore, mercury may be present in trace amounts in digital cameras, laptop computers, and flat-panel displays. However, its use in batteries has been declining for some time, and today all that remains are cells, which power very small electronic devices like watches, toys, and cameras. Compounds containing hexavalent chromium are frequently employed in the electroplating and metal treatment processes. Although it is not frequently used in electronic components directly, hexavalent chromium is frequently found as a coating on the different fittings, and other metal parts used in many items. Both consumer products and those intended for use in tougher situations may contain it since it can be used to protect steel parts against corrosion as well as screws and nuts. Additionally, it is used as a corrosion-resistant substance to safeguard the carbon steel cooling systems in absorption refrigerators.

CHAPTER 6

Hazardous Waste Management

CONTENTS

6.1 INTRODUCTION

Direct and indirect exposure to hazardous wastes has been linked to an increase in the incidence of teratogenic, mutagenic, and carcinogenic effects, reproductive system damage, respiratory effects, brain and nervous system effects, and many other less severe effects. Environmental influences were frequently more obvious and understood. Most people are aware of the aesthetic impacts of improper handling of hazardous waste through the middle of the 20th century. The emergence of the NIMBY phenomenon is a prime example of the public's repulsion for to hazardous materials in general and hazardous wastes in particular (Hermansson, 2007). As property values are affected, jobs are produced or gone, and public managers are under growing pressure to create solutions that fix all problems and resolve all issues, the aesthetic concerns swiftly turn into economic difficulties. Governments, regulated companies and facilities, and the general public all grew very concerned about facility siting. By the 1980s, there was such widespread and vocal resistance to hazardous waste facilities that even locations with stellar operating records threatened closure for no other reason. Environmental equity issues often resulting from the actual or proposed placement of hazardous materials/waste facilities in underprivileged or minority neighbourhoods rose to the top of the urban agenda. By the late 1990s, a specific EPA Environmental Justice programme had arisen to focus on the equity issues. Nearly every aspect of American society has been impacted by the mismanagement and economics of hazardous waste.

6.2 EPACTS

The EPA estimated that 51 million tonnes of hazardous garbage were produced annually by 1979. Governments i.e., local, state, and federal, began making attempts to coerce and force offenders, in various ways, to clean up contaminated sites, stop dumping, stop generating, treat the waste properly, recycle, reclaim, or destroy it as the outrages of hazardous waste mismanagement were thrust upon an ignorant public. Through some actions, people and governments pleaded for relief in the courts. The most frequent defense in the area of environmental law is annoyance. The Clean Air Act of 1970 aimed to reduce the causes of air pollution and enhance the quality of the surrounding air. Little was done to improve the handling of hazardous waste by the 1970 Act. In fact, the industry has built emission control equipment to fulfill CAA-required National Ambient Air Quality Standards and State Implementation Plans, which has resulted in the production of significant amounts of hazardous and toxic material that must be handled as hazardous waste. Only seven compounds had National Emission Standards for Hazardous Air Pollutants (NESHAPS) established by 1990, as required by Section 112 of the Act. By November 2000, EPA had to publish guidelines for 189 more NESHAPS under the 1990 CAA Amendments. An Environmental Impact Statement or EIS had to be prepared for any project that potentially had an impact on the environment under the National Environmental Policy Act of 1970. Without a doubt, NEPA encouraged some improvement in hazardous waste management by numerous agencies. Many cases have been brought against the agencies, but none of them directly compelled cleanup or adherence to environmental law. The main points of the lawsuits have been that either no EIS was done, the EIS was insufficient, or the environmental evaluation was insufficient. Up until recently, the Departments of Defense and Energy mostly disregarded NEPA laws. The Federal Insecticide, Fungicide, and Rodenticide Act-(1972), as modified, controls how pesticides are stored and disposed of. It also mandates that pesticide products bear accurate and detailed labels. Pesticide storage and disposal regulations were published by the EPA in May 1974, and other restrictions that were suggested but never adopted. These rules specified the proper circumstances for soil injection, incineration, and other methods of disposing of pesticide wastes.

6.3 ACTS CONTINUED

In order to prevent or restrict the dumping of items that would have negative consequences, the Marine Protection Research and Sanctuaries Act or MPRSA (1972), often known as the Ocean Dumping Act, controls the dumping of materials at sea (Harvey, 1983). However, because of uncertainties in the Act's original version, it was unclear whether Congress intended to outlaw all ocean dumping or to permit it only in situations where it is the best disposal option and won't adversely harm the marine ecosystem. Congress and the EPA's approval of the ocean dumping of POTW sludge along the East Coast increased this ambiguity. Heavy metals from industrial sources may be mixed into municipal sludge, especially that produced by older cities in the East, on the collection systems. Municipal sludge and industrial garbage were no longer allowed to be dumped in the ocean after December 31, 1991, due to the Ocean Dumping Ban Act amendment to MPRSA that was passed in 1988. Similar to the Clean Air Act, the Clean Water Act's implementation led to the planned removal of waste materials from municipal and industrial wastewater streams, but it also brought up new problems with waste management. Sludge and leftovers from pretreatment and treatment systems were first not regulated and frequently handled incorrectly. Pretreatment requirements must be met by industrial establishments that discharge into collection systems that serve publicly owned treatment works (POTWs). As a result, the CWA regulates and caps hazardous waste emissions into surface waterways as well as POTWs. The CAA regulates hazardous waste emissions into the atmosphere. The MPRSA restricts the disposal of hazardous garbage at sea. By the middle of the 1970s, Congress had theoretically and legally moved the entire burden of final disposal to the land through these many acts. Through two key provisions, the SDWA of 1974 had the potential to significantly protect underground sources of drinking water. The EPA was given the authority to identify specific locations as having aquifers that supply all of the area's water and pose a serious risk to the public's health if contaminated. No government support may be given for any project in the region that the EPA considers may contaminate the aquifer once the area has been so identified.

To prevent contaminated aquifers from being used, the Act regulates underground injection. Most people assume that underground injection is carried out via a deep, high-pressure pumped injection well. Underground injection is the subsurface emplacement of fluid through a well where the depth is greater than its width, according to the 40 CFR 144 standard. The five classifications of injection wells included by this definition include those used to dispose of hazardous wastes among other things. The EPA declared that all states must have underground injection control programmes, and in the event that a state does not, the EPA is required to implement the programme. The programme advanced slowly in the early 1980s because the EPA missed the 1979 deadline for these declarations. On July 29, 1998, the concept of Class V shallow injection wells was proposed. The Toxic Substances Control Act or TSCA of 1976 gives the EPA the power to demand testing of chemicals entering the environment, both new and old, and to regulate them as necessary. It can also be utilized to control the advancement of genetic engineering and biotechnology. The TSCA gives the EPA administrator the authority to impose limitations on the manufacture, transfer, use, and disposal of dangerous substances. The TSCA was the first piece of legislation where Congress specifically named a substance and called for its control. Section 2605 instructed the EPA to gradually stop producing and using polychlorinated biphenyls in accordance with a legally prescribed schedule. The 1965 act's solid waste management programmes were continued and expanded by RCRA Subtitle D, but the Act's and its implementing programmes' provisions for managing hazardous waste were comparatively lax. Amendments made in 1980 to Subtitle C, the legal foundation for the federal hazardous waste management programme, enhanced it a little, but the Hazardous and Solid Waste Amendments of 1984 made a bigger impact.

By placing regulatory requirements on generators, transporters of hazardous waste and operators of treatment, storage, and disposal facilities, Congress aimed to provide cradle-to-grave management of hazardous waste through RCRA. The focus must be on RCRA Subtitle C, which is the statutory basis for federal regulation of the control and management of hazardous waste in the United States and its territories. The equally significant issue of inactive and abandoned hazardous waste sites was not addressed by the RCRA. The CERCLA of 1980, sometimes known as Superfund, contains legislation that establishes remedies and assigns responsibilities for fixing issues at these sites (Sun et al., 2022). When the responsible parties cannot be located, are unable to or unwilling to complete the required cleanup, the Comprehensive Environmental Response, Compensation, and Liability Act or CERCLA provides government money for response and site remediation. The EPA may subsequently file a lawsuit against those found to be at fault in order to recover the money used for remediation. The Superfund Amendments and Reauthorization Act-(1986) significantly altered CERCLA. In addition to extending Superfund money and authority (SARA) significantly expanded CERCLA's the scope of CERCLA. The Secretary of Labor was mandated by Title I of SARA to publish workplace health and safety standards for those who handle hazardous waste. The Disaster Planning and Community Right-to-Know Act, or Title III of SARA, imposed emergency planning requirements on states and local governments and mandated community right-to-know and toxic discharge reporting. In order to require states and communities to create and implement plans for chemical disasters like the disaster that wreaked havoc in Bhopal, India, information regarding chemical use and storage was made available to communities. The 1975 Hazardous Materials Transportation Act encompasses the transportation-related elements of hazardous waste management by allowing the regulation of hazardous material marking, labeling, and packaging. Congress recognized in RCRA the potential for regulatory overlap between the EPA and the Department of Transportation (DOT), and it gave EPA specific instructions to work with DOT to coordinate all transportation-related hazardous waste management regulations. The Hazardous Materials Transportation Uniform

Safety Act of 1990 significantly altered the HMTA. To bring American practice in line with international norms, the 1990 changes were enacted as a general revamp of the packaging requirements.

6.4 HAZARDOUS WASTE

State and local governments made a variety of early attempts to control hazardous waste, ranging from ineffective to non-existent. Early federal legislation aimed to financially support state and local governments, offer them advice and support on a moral level, and encourage them to take positions and actions that were more effective. The most prevalent approach was federal grant-in-aid. The award typically gave start-up money to hire, train, and equip state and local organizations so they could introduce specific initiatives. Federal laws, rules, and policies have increasingly linked financing to performance levels. Federal programmes were delegated to the states, and financing was based on the success of recurrent evaluations of programme achievements. EPA continues to oversee state programmes and delegate authority under RCRA. The majority of states have chosen delegation, and several have implemented RCRA rules. Early legislation and policy regarding environmental management and the cleanup of hazardous waste sites were influenced by concerns about states' rights and police powers. The federal law passed by Congress and carried out by federal agencies was based on the generally held belief that environmental issues were best handled at the state and local levels. The state and municipal authorities were typically underfunded and constrained by a lack of personnel and resources, notwithstanding the clear exceptions. Although similarly constrained, EPA and its predecessor agencies were able to concentrate personnel and laboratory capacity on certain issues. The Congress funded technical assistance capability in the federal agencies in an effort to aid state and local governments in the creation of decentralized programmes. Programs offering technical support were frequently beneficial in locating sources, defining problems, and assessing their impacts, but they were less effective in obtaining prevention and remediation of hazardous waste issues.

Early legislation, initiatives, and policies frequently failed to achieve the objective protection of the environment and of public health. The public and activists called for swift and forceful action to regulate hazardous wastes. By 1994, congressional and federal agency engagement in hazardous waste management and environmental management in general, in general, had grown to such an extent, breadth, and depth that political and public opinion had shifted against these regulation initiatives (Beierle, 2002). Election campaigns in 1994 focused heavily on unfunded requirements imposed by the federal government, and the new Congress pledged to examine many of these initiatives. Individual members of Congress frequently participated in campaigns to halt the expansion or building of such facilities. However, except from for the 1990 CAA Amendments, Congress accomplished nothing in the 1990s to strengthen, weaken, or otherwise alter environmental statutes. The majority of laws, rules, and policies now include procedures that require public hearings and other similar forums. No significant environmental decision is today accepted as routinely being made at any level of government without extensive public engagement, typically through a hearing procedure. These hearings have frequently had a significant impact on how big problems are resolved. Several government agencies have been identified as some of the worst violators of the laws, rules, and policies governing the management of hazardous waste in recent years. President Carter issued an executive order requiring federal agencies to abide by the country's environmental laws in 1978, but it had little impact. Congress passed CERCLA in 1980, but excluded federal government buildings. Federal agencies were not subject to Superfund regulations until 1986. The true nature of the hazardous waste management procedures used by the Departments of Energy and Defense during this time are is now coming to light. These practices were concealed beneath the guise of national security. According to the Departments of Energy and Defense authorities, there may be no way to clean up some of the sites that have been seriously damaged by these procedures. Public forums, hearings, and meetings were useless in several cases, and administrative methods for managing hazardous waste actually worked to hide the severity of the issues.

Similar to the changes previously mentioned, approaches to the definition of hazardous wastes, their identification, impacts, and solutions have advanced. Many incidents where buildings burned down or exploded in the early past released large volumes of heat energy, poisonous gases, and particles. Many acts of burning were confirmed, and many more were suspected. Some locations released hazardous fumes without igniting or exploding. These occurrences were frequently spectacular, frequently dangerous to the health and/or safety of people, and always frightful to the general population. At the time, it was known that the groundwater resource provided drinking water for more than 50% of the population and served as the water supply for 30% of home water systems. Many soluble and leachable inorganic and organic contaminants, many of which are poisonous, were contaminating aquifers. The effects of groundwater contamination may last for decades or centuries, in contrast to the discharge of harmful air pollutants and the heat energy from fires and explosions, which can be quantified in days, hours, minutes, or even seconds. The attention of Congress and the regulatory organizations started to change to groundwater pollution prevention and contaminated aquifer remediation. For many years, many East Coast cities barged their industrial waste-filled sewage sludge to coastal dumping grounds. Ocean dumping of hazardous wastes became a common alternative in the 1970s as a result of the CAA and CWA regulatory programmes and their' ever-tighter limits as well as the disappearance of land disposal sites.

The majority of hazardous waste was thus diverted onto the land or below the surface of the land as a result of the legislation and implementation of the 1970s. Land farming and surface impoundments were proposed as treatments, and deep well injection systems and landfills were accepted disposal methods. The impoundments and land treatment facilities, in actuality, provided only minimal treatment, allowing their toxic elements to enter the groundwater resource. Many professionals questioned the environmental safety of deep well disposal because landfills also rendered leachable and liquid toxic elements accessible to groundwater. When Congress passed HSWA in 1984, these factors took precedence. There were many different types of treatment technologies available by the early 1990s, and there was a good chance that the amount of hazardous waste controlled for land treatment and disposal would be significantly reduced. However, widespread withdrawal of RCRA permit applications, cancellation of construction plans and contracts, and a noticeable reluctance toward corporate involvement in new facilities were the results of increasingly active activism and public opposition to the siting of new treatment and destruction facilities. Americans have been seen to be several steps behind Western Europeans in the development of hazardous waste management(Lampe & Gazda, 1995). It has long been believed that there isn't enough land in Europe for land disposal to become the preferred alternative. Early on, incinerating waste became the preferred method, and it still is today, but many methods of disposing of land are used. In Britain, for instance, 85 percent of hazardous waste is dumped in landfills. Land disposal is used in Denmark, France, and the former West Germany, however, these nations obviously place an emphasis on the procedures of treatment and destruction.

To avoid purchasing expensive scrubbers for land-based incinerators, the West German chemical sector started burning used chlorinated hydrocarbons in the ocean. High-temperature incineration efficiently eliminates the wastes, but incinerators lack room for equipment to clean stack gases, and the machines release hydrochloric acid. The HCl released by incinerators was expected to be neutralized by the seawater capacity as a buffer. Europeans lost interest in this technique and compelled the phase-out of ocean incineration in international accords. A U.S. business made financial and resource commitments to an ocean incineration project. Despite the lack of available space for land disposal, both European and American land treatment and disposal practices involved some costly mistakes. In Britain, co-disposal, the intentional blending of hazardous wastes with regular municipal wastes in permeable landfills was one of the earliest pillars of the toxic waste plan. In Holland, where much of the land is below sea level, fill is frequently necessary before any land construction can begin. Solid wastes containing dangerous chemicals were used as fill material for more than 40 years. Authorities believe that there may be up to 8 million metric tonnes of dangerous chemical waste buried there. According to government estimates, there are currently close to 5000 waste leakage sites. Particularly in Denmark, incineration in integrated power and heating facilities, recycling, and waste minimization are all highly valued. Waste disposal in landfills has been cut to 20%. Landfilling garbage that could have been burned for energy recovery has been prohibited since 1997. Pump-and-treat techniques have been widely utilized to clean up contaminated aquifers, and they appear to have worked at some locations in Denmark.

6.5 POLLUTION SITES

As the Soviet Union fell apart, thousands of polluted sites were left behind by Russian military troops as they evacuated their garrisons, airfields, seaports, and training facilities in Central and Eastern Europe. These include artillery ranges, practice fields, tanks and storage facilities both above and below ground, fuelling locations, and maintenance and repair areas. The sites are contaminated with a variety of contaminants, including hydrocarbons, heavy metals, acids, PCBs, PAHs, solvents, ordnance, and biological agents. The host countries must now identify ways and means of responding to the possible threats because the occupiers have not accepted financial responsibility for remediating these sites. Latvia, Estonia, and Lithuania all have comparable situations. Due to Russian naval activities, the Baltic countries have also had to clean up their dirty harbors, bays, estuaries, as well as the open Baltic Sea. There were many fuels, munitions, and chemical wastes in the ports, bays, gasoline stations, communications hubs, ammo depots, and rocket bases. The remains of sunken ships are another source of pollution in the marine areas. The North Atlantic Treaty Organization (NATO) provided funding to academics from the Environmental Technology Program at the East Campus of Arizona State University in 1997 to perform cooperative studies at missile sites close to St. Petersburg and Moscow, Russia, in an effort to better comprehend this Cold War legacy (Nath & Marinova, 2000). The former military stations appeared to be the site of widespread radioactive waste dumping, according to their preliminary field studies; nevertheless, Russian government officials are reluctant to permit a complete examination and analysis of the issues. The investigations have been put on hold as a result of this unofficial resistance. According to the ASU professors, the prospective problem sites may equal the U.S. Department of Energy weapons facilities in terms of the volume of wastes generated and the geographic areas in need of remediation. Due to the remoteness of the sites, the severity of the environment, and the limited fieldwork seasons, the contaminants, notably petroleum, lubricants, and PCBs, present particularly challenging cleanup difficulties.

Provincial hazardous waste management programmes in Canada are somewhat similar to those in the United States. The provinces have the primary responsibility for regulatory programmes, but the federal government does have limited interprovincial power. With contemporary, integrated treatment facilities now operational or in the works, Canada's capacity for off-site treatment and hazardous waste volume reduction is expanding. One example is the Bovar plant in Swan Hills, Alberta, which just started accepting imports of hazardous garbage. As a result, there are no stringent laws in place that govern the treatment and disposal of hazardous wastes. Nine kinds of hazardous compounds are listed in the Canadian Environmental Protection Act as being either restricted or subject to controlled release. A Leachate Extraction Procedure resembling the Resource Conservation and Recovery Act (RCRA) has recently been added. The federal government announced fresh regulatory action in April 1999 with the goal of reducing dichloromethane discharges by 63 percent by 2002 and 85 percent by 2007 as well as hexavalent chromium releases by 75 percent by 2005. Toxic material discharge practically goes unchecked in Mexico's rural communities. Hazardous wastes from the United States were imported as a result of the lack of restrictions, but RCRA requires exporters to notify EPA in detail and secure the receiving consent. The management of hazardous waste has come into sharp focus as a result of the expanding industrial activity along the U.S.-Mexico border. To assure safe treatment and/or disposal and to comply with duty-free requirements of U.S. and Mexican customs rules, the maquiladora businesses are said to import their hazardous wastes from their Mexican facilities into the U.S. The import of hazardous waste into the US is also governed by RCRA (Lara-Valencia, Harlow, Lemos, & Denman, 2009). The environmental effects of these operations in some of the industrialized border areas are severe, and responsibility for hazardous wastes produced by maquilas has not proven to be sufficient. The analytical chemist may have to deal with the definition of hazardous waste from a variety of angles more than others. He or she might be asked to define hazardous waste in terms that make it possible to carry out analytical determinations and/or screening procedures quickly, affordably, and thoroughly enough to prevent loopholes. He or she might be asked to create analytical or screening processes, or to choose the best alternative among numerous processes. The chemist may determine that it is important to set up a lab to handle the analytical needs of a certain source as effectively as possible. He or she might work on the control of a production or treatment process where wastes could change from hazardous to non-hazardous due to controllable factors.

6.6 CHEMICAL ANALYSIS

The chemist is especially worried about the security of analytical techniques. The chemist must develop processes that enable the decision to be made without exposing the analyst and/or others to risks when screening techniques are used for decision- making in the field or onsite. He or she is expected to describe hazardous waste in chemical terms that are both clear enough to survive the rigors of a courtroom and sufficiently complex to allow needed testing to be safely carried out in the field by semi-skilled employees. This contrast is accentuated by the fact that many of the analytical techniques outlined in SW 846 are extremely complicated, involving advanced equipment and processes that are typically opaque to courts, the media, and the general public. Expert witnesses or witnesses who testify regarding chemicals typically include analytical chemists. The criterion used to determine whether a sample is hazardous or non-hazardous is the regulatory definition, which is based on the law. Ambiguity or an overly complicated definition can make the testimony difficult or impossible for non-lawyers to understand and therefore hinder or prevent credible enforcement actions. The conclusions in such conditions can, of course, be extremely important. The biological effects of exposing live cells to hazardous wastes are addressed by the roles of the life scientist and the health professional in hazardous waste management. The effects of exposure on nonhuman cells, which operate as indicator organisms, are of particular interest to life scientists. The hazardous waste ingredients that produce the disease or genetic damage, as well as the method or means by which the waste constituent affects the human target, are all of interest to the health professional. The life scientist may be requested to create or enhance bioassay techniques that will be used to create or change exposure criteria or to assess a shipment or category of garbage in comparison to predetermined standards. He or she might be asked to assess bioaccumulation rates and/or effects of toxic components in hazardous wastes, assess a certain waste treatment method or recommend a bioremediation method that might be expected to satisfy a cleanup requirement. The task of converting the data from life scientists about nonhuman exposure to human exposure standards may be given to the health professional. Other duties can include determining a threshold level based on measured exposure levels and morbidity statistics or offering expert testimony on cause and effect in exposure cases.

It is expected that the life scientist and the health professional will define hazardous waste or assess a waste product in accordance with a recognised recognized life science or health standard. As before, it is uncommon for conditions to allow for in-depth, sophisticated, or real-time scientific examinations of a waste cargo or a batch of waste that has been collected. The same as before, the problem is to define hazardous waste in such a way that the method is quick and easy while yet protecting the environment and human health. In a broader sense, environmentalists work to safeguard the environment from hazardous waste through activism, advocacy, legislation drafting, and education. Finding a practical definition of hazardous waste is important and problematic for environmentalists. The link between DDT and bioaccumulation, egg shell thinning, and potential extinction of significant species led to the resolution of the DDT problem (Nakamaru, Iwasa, & Nakanishi, 2003). There aren't many of these options among the countless wastes, components, combinations, and concentrations that could be discharged or could happen after release. Environmentalist-proposed criteria would undoubtedly face opposition from special interests that require confirmation of direct cause-and-effect relationships. The real process of establishing a substance's environmental impact may be masked by a number of sub-processes and take years to complete. Legislators, attorneys, administrators, and diplomats are worried about the workability of the definition, maybe without making a major distinction from what was previously stated. Regulations must be founded on statutes. Regulations need to be clear and enforceable. The public must be protected against exposure to unsafe concentrations or amounts of hazardous waste, and regulatory agency administrators must have the statutory, regulatory, and financial resources necessary to do so.

Sorting out each participating government's idea of what wastes are being addressed is a constant concern in diplomatic efforts to secure international and/or regional hazardous waste management agreements and treaties. Exact The exact point is made by the United Nations Environmental Programme. Offsite recycling is frequently used to reduce waste, but exports of waste intended for recovery that are poorly defined and specified let in unlawful activity. Different countries may use quite complex criteria, while others may merely use the justification that any chemical that is disposed of is a hazardous waste. Even after 20 years of cradle-to-grave waste management, it is still challenging to estimate the proportional contributions of different source types. Because hazardous pollutants contained in wastewater discharges were not included in the 1997 data, the overall amount created appeared to have decreased significantly. For instance, it is debatable whether or not wastes that are exhumed from a remediation site and transported to an authorized treatment, storage, and disposal facility qualify as freshly created waste.

Even though other than manufacturing facilities produce a lot of hazardous waste, the TRI gives some insight on the types and relative volume of sources of hazardous waste releases to the environment. It is possible to categorize atmospheric discharges as regulated or uncontrolled. In most areas, open garbage burning is no longer permitted, yet it still happens on a legal and illegal basis. The production, sale, and burning of hazardous waste fuel as well as the disposal of hazardous wastes by authorized combustion facilities are all subject to RCRA regulations. The overarching goal of these regulations is to ensure that hazardous waste is burned under conditions that will assure its destruction and that any residues are appropriately caught and controlled. Early in the 1990s, the practice of combining hazardous waste with other fuels and then burning it in boilers and industrial furnaces became and remains a very divisive topic. Chlorofluorocarbons are also released by vehicular refrigeration systems. Operations including bulk disposal and land farming make volatiles available for evaporation while leaving the heavier components to soak into the soil. By purpose or by accident, impoundments for hazardous waste leak its volatiles into the atmosphere. In order to improve vaporization, hazardous wastes containing volatiles are nevertheless thinly scattered on the soil surface by land treatment facilities. Digester gases containing volatiles from sludge are frequently released by sewage treatment plants. In publicly owned treatment facilities, activated sludge and aeration basins, trickling filters, and holding ponds are particularly successful at removing volatiles from sewage and, unless they are caught, releasing them into the atmosphere. Similar to how hazardous waste tank storage facilities release the lighter volatiles as the tank's vapor pressure changes.

6.7 EMISSIONS

Chemical manufacture, primary metals, paper, and food processing were the industries with the highest amounts of chemical emissions to surface waterways in 1997, as well as their proportions of the total releases to surface water. Prior to discharge to publicly owned sewage, hazardous elements in industrial waste streams are required to be removed or destroyed by pretreatment regulations, standards, and codes. Surface waters get large volumes of the liquid component of sewage discharge. Chemicals in sewage may escape removal in sewage treatment plants by passing through or interfering with treatment processes. Chemicals extracted by sewage treatment procedures may be released into the environment as sewage sludge or as air emissions. The EPA has increased its pressure on state and local regulatory organizations to strengthen pretreatment ordinances and regulations and to increase compliance with pretreatment standards. If surface water and urban air quality goals are to be met, industrial waste discharges to POTWs must undergo the proper pretreatment. The Marine Protection Research and Sanctuaries Act has restricted ocean dumping, but substantial amounts of hazardous material have nonetheless found their way into the marine ecosystem. A U.S. district court decision arising from a City of New York lawsuit, which claimed that the EPA had improperly executed the legislation, cast doubt on the Act's execution (Freeman & Vermeule, 2007). Despite fierce protests from environmentalists and the general public, permitted dumping of hazardous waste and ocean disposal of sludge were both continued. Reducing the risks to human health and the environmental effects of land treatment and disposal practises practices has been the main goal of the ongoing evolution, which includes RCRA, the Hazardous and Solid Waste Amendments, and the site remediation rules and strategies of the Comprehensive Environmental Response, Compensation, and Liability Act. Nevertheless, substantial amounts of hazardous garbage are still being handled, dumped, and/or buried in the ground. A remediation method that keeps polluted soil or groundwater in situ while natural attenuation lessens the contaminant's mobility or toxicity has recently received tentative approval from the EPA in the form of a policy directive. Land disposal accounted for only 9% of the national total in 1991, according to the EPA, whereas aqueous treatment units handled 76% of it. Land disposal methods made up 12.3 percent of all hazardous waste handling in 1995, according to the EPA. According to the 1997 TRI data, about 17% of SIC 20-39 manufacturing hazardous wastes were disposed of on land. However, it is clear that there has been a change away from land disposal.

Despite trends, large amounts of hazardous waste are still being released into the air, surface water, land, and subsurface. It is now crucial to comprehend what happens to the waste elements after these dispositions. Saying that waste elements have an influence based on the concentrations present at the site and time of exposure, the extent to which the concentrations grow or drop during the exposure, and the duration of the exposure is oversimplifying, but conceptually relevant. Rates of dispersion, rates of accumulation and decay, and residency times all have significance in light of this idea. The idea can be expanded to include that the effects of a release of hazardous waste on human health and the environment may be strongly influenced by the medium to which it is released and by the forces that operate upon the waste after release. Although the earth's atmosphere reaches several hundred miles above its surface, the majority of it, approximately 95% of it, is concentrated in a layer that is only about 12 miles thick. The troposphere is the bottom portion of the 12-mile layer that contains the majority of the air mass. In At the poles, the troposphere is about 5 mi thick; at the equator, it is roughly 10 mi thick. The troposphere makes up around 75% of the overall mass of the atmosphere due to gravity's pull and gases' ability to be compressed. The troposphere's behavior and the forces operating on it control how pollutants released into the atmosphere behave. Tropospheric air is warmed by the sun's energy through convection, conduction, and radiation.

Nitrogen and oxygen make up the majority of the gases that make up air. Particulates and water vapor are also present in the air in its natural condition. The hazardous wastes that are discharged into the atmosphere can and do undergo an infinite number of chemical reactions due to these components, heat energy, and photochemical activity (Batchelor, 2006). While the formation of carbon monoxide and carbon dioxide in combustion processes, the conversion of released nitric oxide to nitrogen dioxide and nitric acid in the atmosphere, and the roles of sulfur releases in the formation of sulfur oxides and acid rain are all well-known, other chemical reactions in the atmosphere are of concern to the hazardous waste management. Further physicochemical alterations to the atmosphere are possible as a result of these discharges. Releases to the unrestricted atmosphere are more likely than releases to the ground or water to disperse quickly. When substances are released into the atmosphere, concentrations at the point and time of exposure are often much lower than at the site of release. Following its emission, air pollution concentrations typically decrease over time. Chronic rather than acute consequences are more often the result of these traits. Releases Emissions entrapped by an inversion may quickly build up to harmful amounts, may continue throughout the inversion, and may have an immediate negative impact on human health. Releases Emissions in the home and at work may have both short-term and long-term health effects.

6.8 POLLUTANTS

Polychlorinated biphenyls and the banned pesticide DDT have been found in the fish and fauna of the Great Lakes after moving from polluted soils into the air. Acetone, toluene, and xylene are among the most commonly reported TRI chemicals that are released and react with other substances to produce ozone, a less harmful air pollutant. Methanol, which has a 1.11 vapor density, tends to stay at or near the surface. It poses a fire risk unless it is removed. 1,1,1-trichloroethane that has been released is extremely persistent and can move to the upper atmosphere, where it plays a crucial role in the destruction of the ozone layer that protects the stratosphere. When found at large concentrations, ammonia is a very harmful irritant and asphyxiant, yet it dissipates quickly into the atmosphere. It is commonly known that acid rain components spread over state lines, borders, and even continents, causing fallout on land and ocean surfaces. It is recognised recognized that waste CFCs that deplete the ozone go from the surface of the earth to the furthest reaches of the atmosphere (Blake, 1988). Vanadium, manganese, and lead are among the heavy metals that are carried across and around the north pole from industrialized regions of Europe and Russia. The 1986 Chernobyl Chornobyl nuclear power plant disaster left radioactive waste that was transported and dispersed throughout Yugoslavia, France, Italy, Germany, and Scandinavia. Due to the restricting effect of the stream channels, the movement of released chemical constituents in surface streams is slightly more confined than in the atmosphere. The water-air barrier causes the emission of volatile organic molecules. Higher molecular weight organics and soluble inorganics can be transferred from running streams to the groundwater. Pollutants may be transferred through downstream diversions to the surface of the land or to home or commercial water sources. Inorganics may be precipitated and dumped on the ground, concentrated and returned to the source stream, or stripped of their organic components by aeration procedures and vented to the atmosphere.

There are groups of chemical waste elements in surface waters that have a negative impact on the environment. They consist of heavy metals, their compounds, and synthetic organic materials that cannot decompose. As they go up the food chain, these compounds may accumulate to pathological or deadly levels, even in very little amounts. Because they ultimately flow into the delicate habitats of the coastal waters, the movement of hazardous waste elements in surface streams is significant. The concentration of releases to impounded surface waters may be substantially greater, and they may also transport dangerous elements to the atmosphere or to groundwater. Artificially aerated surface waters also release volatile organic compounds (VOCs) into the environment. Many of the ponds were built with the intention of eliminating wastewater by seepage, evaporation, or both. Using typical leakage coefficients, the EPA projected that these impoundments have allowed more than 100 billion gallons of industrial effluent to reach the groundwater system. The United States' source of freshwater in the US is largely made up of groundwater. The population depends on groundwater for household purposes to the extent of more than 50%. Water is transported through and stored in the earth's crust's water-bearing formations. Water enters these formations from the ground's surface or from bodies of surface water, then moves slowly until it either percolates into deeper formations or returns to the surface as a result of natural flow, plant growth, or human activity. When surface run-off is low or absent, groundwater that enters stream channels as base flow helps to maintain stream flow.

Almost all groundwater starts out as surface water. Rainfall, stream flow, lakes, and reservoirs are the main sources of natural recharge. In addition, man-made activities such as excessive irrigation, but more specifically, the disposal of liquid waste in pits, impoundments, landfills, and other land applications, as well as the leaching of solid or semisolid hazardous wastes, result in additional contributions, which are referred to as artificial recharge. One or more of a number of techniques may be used to transport hazardous waste components underground. Volatile pollutants' vapors can diffuse through voids above the water table, where they can dissolve in water that is trapped inside or seeps through the holes. Depending on their densities, petroleum products and chlorinated solvents that don't mix well with water may run on the surface of the groundwater or sink. Contaminants may sorb onto colloids or naturally occurring organics and travel through them. Metals can precipitate and move in a similar way. These mechanisms might alter or fluctuate as soil particles and spaces get saturated or as the pH of the subsurface environment shifts. The amount of wastewater being treated, stored, or disposed of in unsealed surface impoundments has significantly decreased as a result of the implementation of HSWA land disposal limitations. The EPA calculated that in 1989, surface impoundments were receiving about 31 million gallons of liquid hazardous waste. Contaminated sites that need to be cleaned up, spills, and sources that are not regulated continue to leak hazardous waste components into the atmosphere, surface waterways, and groundwater. The argument is that a significant amount of hazardous waste that was dumped on land has evaporated into the atmosphere, run-off into surface waters, and percolated into groundwater. The groundwater contamination may be highly concentrated, relatively confined, and persistent for decades or millennia while the air and surface water emissions are lost to natural processes or mixed with other releases. Although these amounts are decreasing, the ongoing discharges, together with the materials that have already been discharged, have poisoned many aquifers and continue to do so. Many groundwater supplies have also been damaged or destroyed.

Trichloroethylene and tetrachloroethylene have historically been the most common hazardous waste releases to the land surface, landfills, and surface impoundments (Reinhart, 1993). Both have been used extensively as degreasers and cleaning solvents, and up until strict regulations were implemented, they were disposed of using the easiest manner at hand. When bacteria from sewage or septic tank leachate were present, EPA scientists found that vinyl chloride could frequently be found in soil samples where TCE, PCE, and other chlorinated chemicals had been found. It was quickly determined that dechlorination of trichloroethylene (TCE) and tetrachloroethylene (PCE) TCE and PCE can be anticipated to advance be converted to vinyl chloride (VC) under anaerobic circumstances. VC is a recognised recognized synthetic carcinogen that is ranked number 4 on the CERCLA List of Priority Hazardous Substances and number 21 on the list of the most frequently found CEPs at hazardous waste sites. Mercury has been emitted into the atmosphere and into surface and groundwater through a wide range of industrial operations, despite an increase in the number of controlled sources.

Mercury can exist as a solid, liquid, or vapor as well as in elemental form and as any of a large number of different organic and inorganic compounds. It easily undergoes biotransformation, is ingested by plants, and is concentrated through food chains. While the mobility and activity of other hazardous wastes are important considerations in their management, these wastes may be more or less quickly converted, mobile, and/or dangerous to the environment and human health. These elements significantly increase the complexity of the imposed control systems. All cells must allow for and promote a range of chemical reactions in order to function and maintain themselves. When a foreign chemical is introduced into a cell, it may interfere with one or more of these biological processes, impairing cell viability or function. All chemicals are poisonous, but how hazardous they are depending depends on the concentration, entry point, and duration of exposure. The study of how certain substances harm live cells and entire organisms is known as toxicology. Such investigations are done to find out what cells are impacted by the chemical, what cell functions are compromised, how easily the chemical enters the organism, how it acts inside the organism, and how quickly it is expelled from the body. The level of chemical exposure deemed acceptable for people or other organisms is determined through the use of a risk assessment procedure. Then, risk-based standards are used to rationalize risk-based exposure limits. The technology-based standard is an alternative to exposure limits, and it aims to reduce exposure by imposing control measures. Significant progress has been made in toxicology and the research techniques used by toxicologists recently. However, the basis for quantitative toxicity assessments is not sufficiently supported by current toxicological information for a large number of compounds.

Many methods of delivery can make chemical waste elements immediately or chronically toxic to plants or animals. Wastes that are phytotoxic can harm plants if they are present in the soil, air, or irrigation water. A decrease in the ability to produce chlorophyll, general growth retardation, or a specific chemical interference mechanism can all lead to phytotoxicity. Chemical elements that are immediately harmful to mammals may cause harm when inhaled, consumed, or come into touch with the skin. Acute exposure symptoms typically appear during or soon after exposure to a pollutant at a sufficiently high concentration. Chemicals differ greatly in the concentration needed to elicit these effects. It's possible that information relevant to one specific route of administration won't apply to others. For instance, asbestos dust, which is deadly at extremely low quantities when present in the air, is thought to not constitute a threat to human health when present in water at low concentrations. The classic examples of chronic toxicants include heavy metals and halogenated aromatic chemicals. About 35,000 chemicals have been identified by the U.S. Environmental Protection Agency as either definitely dangerous or possibly harmful to human health. Some of them are carcinogenic, including some heavy metals and specific chemical compounds. Others, like mercury, are mutagenic and have a propensity to cause neurological damage, kidney disease, bone and brain damage, among other issues. The risk brought on by concurrent exposure to more than one of these chemicals is frequently unknown, however, multiple exposures can be additive or synergistic in some circumstances.

A chemical must first enter an organism's cells and tissues in order to have an adverse effect on those cells and tissues. Inhalation, dermal absorption, and ingestion are the three main ways that harmful chemicals enter bodies. Regarding how quickly chemicals traverse the tissues involved in these three exposure routes, there are differences in their absorptive surfaces. Chemicals come into touch with the gastrointestinal tissues during ingestion. The GI tract's primary job is to absorb the food and liquids that are consumed, but it is also capable of absorbing hazardous compounds that are present in the food or water. The degree of absorption typically relies on whether the substance being consumed is hydrophilic or lipophilic. Since the chemical can easily diffuse across the membranes of the cells (GI) tract, lipophilic substances are typically well absorbed. Hydrophilic substances must be carried across by a transport system within the cells because they cannot traverse the cell lining in this manner. The effectiveness of the transport system and the chemical's similarity to commonly transported compounds determine the scope of the transfer. If the substance being consumed is a weak organic acid or basic, it will likely diffuse towards the area of the digestive tract where it is most lipid-soluble and be absorbed there. The polarity of a chemical can vary significantly in these regions of the GI tract since the gastric juice in the stomach is acidic and the intestinal contents are almost neutral. Weak organic acids are more likely to be absorbed by the stomach because they are in their least polar state in the stomach. A weak organic base prefers to be absorbed through the colon since it is in its least polar form there.

6.9 ABSORPTION INTO THE BODY

A chemical must first enter an organism's cells and tissues in order to have an adverse effect on those cells and tissues. Inhalation, dermal absorption, and ingestion are the three main ways that harmful chemicals enter bodies. Regarding how quickly chemicals traverse the tissues involved in these three exposure routes, there are differences in their absorptive surfaces. Chemicals come into touch with the gastrointestinal (GI) tissues during ingestion. The GI tract's primary job is to absorb the food and liquids that are consumed, but it is also capable of absorbing hazardous compounds that are present in the food or water. The degree of absorption typically relies on whether the substance being consumed is hydrophilic or lipophilic. Since the chemical can easily diffuse across the membranes of the cells (GI) tract cells, lipophilic substances are typically well absorbed. Hydrophilic substances must be carried across by a transport system within the cells because they cannot traverse the cell lining in this manner. The effectiveness of the transport system and the chemical's similarity to commonly transported compounds determine the scope of the transfer. If the substance being consumed is a weak organic acid or basic, it will likely diffuse towards the area of the digestive tract where it is most lipid-soluble and be absorbed there. The polarity of a chemical can vary significantly in these regions of the GI tract since the gastric juice in the stomach is acidic and the intestinal contents are almost neutral. Weak organic acids are more likely to be absorbed by the stomach because they are in their least polar state in the stomach. A weak organic base prefers to be absorbed through the colon since it is in its least polar form there.

The interaction of the chemical with gastric or intestinal contents is another crucial factor in determining absorption from the GI tract. A chemical taken in food is frequently not absorbed as effectively as when it is ingested in water since many compounds have a tendency to bond to food. Additionally, some compounds might not survive in the stomach's highly acidic environment, and others might undergo changes brought on by digestive enzymes or intestinal bacteria, resulting in new molecules with different toxicological characteristics. For instance, gut bacteria can convert aromatic nitro groups into potentially cancer-causing aromatic amines. Regardless of the method of absorption, the chemical is transported to the target organ once it has reached the bloodstream. Workers eating sandwiches for lunch in battery manufacturing might consume lead. With long-term consumption of pollutants in water supplies, ingestion becomes more significant. Chemicals come into contact with the lungs during inhalation. Gases or vapors of volatile liquids make up the majority of inhaled substances. Because of the lung's enormous surface area and proximity to blood arteries, absorption there is often very high. Simple diffusion allows gases to pass through the lung membranes, and the rate of absorption is based on how easily the harmful substance dissolves in blood. While the blood flow through the lung limits the rate of absorption of gases with low solubility, the rate and depth of respiration are the only factors affecting the absorption of readily soluble gases. Chemicals can also be breathed as dust or aerosols in solid or liquid form. If lipid-soluble, liquid aerosols will passively diffuse across cell membranes. The chemical composition of the particles has a significant impact on how well they are absorbed. The solubility of the component in lung fluids influences the rate of absorption of particles from the alveoli, with poorly soluble compounds being absorbed at a slower rate than readily soluble compounds.

The alveoli may hold onto small, insoluble particles indefinitely. Larger particles are either swallowed and deposited in the GI tract or they are deposited in the trachea or bronchial areas of the lungs where they can be removed by coughing or sneezing. Particles that are 5 m or larger are typically deposited in the throat or nasal passages, where they are later ejected or ingested (Kublik & Vidgren, 1998). If the lung's defensive mechanisms are overloaded with particles from smoke, coal dust, etc., a persistent effect on the lung may result. In industrial settings, inhalation of air pollutants is likely the most significant pathway for chemical absorption into the body. Over the course of an 8-hour work shift, a worker exposed to 1000 ppm of toluene vapor could be expected to exhibit significant symptoms of respiratory and eye irritation as well as central nervous system depression. This response to toluene exhibits both local and systemic effects, as the substance influenced the CNS after being absorbed into the bloodstream. Some substances lack warning properties when they are in a gaseous or vapor state. For instance, odorless and colorless carbon monoxide can have devastating harmful effects on the unaware victim. The receptor may become less sensitive to other substances. For instance, hydrogen sulfide has a strong rotten egg smell even in small amounts. The olfactory senses are paralyzed in large quantities, though, and the poisonous effect can quickly overtake an exposed person.

Humans may develop long-term chronic health impacts in a variety of circumstances. For instance, persistent intake of sulfur dioxide, one of the more prevalent urban air pollutants, has been conclusively related to chronic bronchitis. Exposure to dust and/or fumes has been linked to emphysema, asbestosis, silicosis, and berylliosis. The thick layer of rough, keratinized epidermal cells blocks the absorption of toxicants through the epidermal layer of the skin and into the bloodstream. Chemicals are absorbed easily through damaged skin. Skin structure varies significantly from one area of the body to another, and these variations also affect dermal absorption. Chemicals can be applied in an oil carrier and then rubbed into the skin to increase their absorption by the skin, which is roughly proportionate to how well they are soluble in lipids. Some lipid-soluble substances can be absorbed by the skin in large enough doses to have an impact on the entire body. For instance, large enough doses of carbon tetrachloride can cause liver damage when absorbed via the skin. The physiological impact that toxic substances have on the exposed species can be used to classify toxic compounds. Although the categories frequently overlap, they can be divided into irritant, asphyxiant, CNS depressive, and systemic toxicant groups.

Chemicals that have local effects at the place of introduction into the body, such as discomfort, erythema, and swelling of the skin, eyes, respiratory tract, or GI tract, are referred to as irritants. Sodium hydroxide dust on perspiration-soaked skin is one illustration. Irritation results from the fluid's pH being swiftly raised over normal. The irritating impact is amplified by mechanical friction, such as the rubbing of shirt collars or cuffs. The result can range from a minor stinging sensation to more severe skin blistering. Ammonia vapors or spray can puncture and burn the nasal passages and throat, irritating the mucous membranes of the respiratory tract. Chemical asphyxiants are substances that deprive host organism cells of oxygen, slowing or stopping metabolism in the process. Simple or mechanical asphyxiants reduce the amount of oxygen in the air to the point where it becomes inhospitable to life. When methane gas replaces oxygen in an enclosed space to the point where the atmosphere's oxygen level drops to under 16 percent, oxygen starvation may result. On the other hand, after being inhaled, carbon monoxide is a gas that chemically binds up the hemoglobin in the blood. Death occurs because the tissues are unable to continue their normal metabolic processes because hemoglobin is unable to deliver oxygen to the cells and carbon dioxide away from them. The nervous system is damaged by the inhalation of organic solvent vapors, anesthetic gases, or narcotics like alcohol or depressants that are introduced to the body. A worker who inhales trichloroethylene vapor during a shift might not be able to drive a car safely due to poor neuromuscular coordination. It is possible to confuse the symptoms of a high blood alcohol concentration with the appearance of inebriation.

6.10 TOXICANTS

Chemical substances known as systemic toxicants manifest their effects significantly on a specific organ system and may do so far from the site of entry. The systemic toxicants and the other groups have a lot in common. For instance, the organic solvent carbon tetrachloride undoubtedly depresses the central nervous system (CNS), is irritating, and might harm the liver or kidneys irreparably (Bergman & Appelgren, 1983). While inhaling mercury vapor does not appear to cause irritation, it seriously damages nerve endings. Chronic mercury vapor inhalation can cause major neurological system disorders. A substance is referred to as a carcinogen or possible carcinogen if it has the capacity to cause the abnormal, excessive, and uncontrolled proliferation of specific cell types or the abnormal division of cells. Lung cancer and mesothelioma have been definitively linked to inhalation of asbestos fibers. A mutagen is a substance that alters or induces mutations in the DNA found in chromosomes. Salmonella sp. experiences these modifications as a result of exposure to formaldehyde vapor. Future generations could be impacted by mutagenic toxins. A toxin known as a teratogen causes physical abnormalities in unborn children.

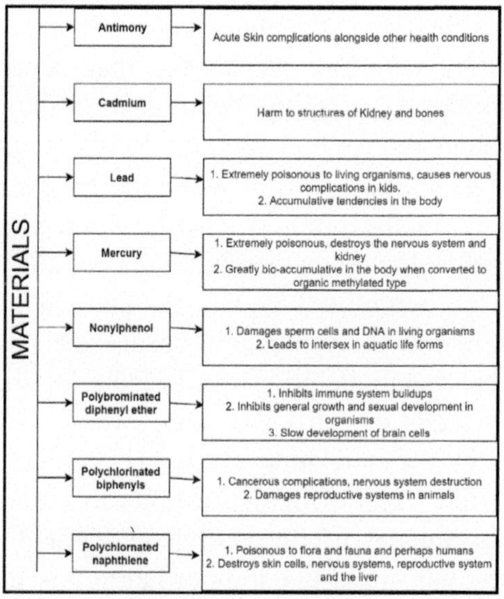

Figure 6.1: An overview of hazardous substances. (Source: Akpan VE, Olukanni DO. Hazardous Waste Management: An African Overview. Recycling. 2020; 5(3):15. https://doi.org/10.3390/recycling5030015)

CHAPTER 7

Environment and Health Impacts

CONTENTS

7.1 INTRODUCTION

As long as there have been large populations to place a strain on the available resources, managing solid waste has been a problem in civilization. In the past, waste from houses and businesses could be disposed of by simply transporting it to unsanitary dumps, where it may be buried, consumed by animals, or burned. To prevent annoyances and issues with public health, household waste and other waste streams needed to be eliminated from the human environment. The larger environment offered an adequate sink for these unfavorable effects on human life. As a result, waste management practices and policies in industrialized nations developed quickly in the second half of the 20th century to ensure that environmental resources are maintained while risks to public and occupational health are kept to a minimum. The main objective of sustainable development has been one of the driving forces in influencing waste policy, along with many other aspects of daily life in society, since the 1980s. Within this broader goal, waste management strategies have evolved to incorporate the social, economic, and environmental aspects of sustainability. In actuality, integrated waste management, the prudent application of a variety of alternatives to establish a generally ideal system of waste management and resource recovery has been equated with sustainable waste management. Breaking the historical connection between waste production and wealth development is the major difficulty facing a contemporary, industrialized nation. The amount of garbage produced per person and wealth have both seen to increase steadily over time, outpacing economic development.

7.2 WASTE DISPOSAL RESEARCH

Although the formulation of estimates for future waste trends is hampered by a lack of data, it is believed that the majority of waste streams will likely increase during the next ten years. It is still highly challenging to increase efficiency in resource use and lessen the negative environmental effects of the flow of waste materials and energy. This is due to issues with prices and acceptance rather than any specific technical limitations. More environmentally friendly waste disposal methods can appear to be prohibitively expensive alternatives, especially if there is no consensus regarding who should pay for those expenses rResearch conducted by the Dutch Environment Ministry in 1997 revealed that the proportion of municipal garbage to household waste might range from 45 to 84 percent (Jha et al., 2008). Recognizing that municipal garbage is a management concept and that the breadth of municipal collection activities in the business sector varies considerably across the nations studied makes it unlikely that statistics and information will always be directly comparable. A minor portion of all solid wastee is MSW. In many respects, MSW is the most important waste stream since it originates from each of us, reflecting our consumption habits, lifestyle choices, and decisions regarding resource recovery. Waste policies typically focus on establishing goals that require nations or local governments to recover or recycle a certain amount of waste, or to divert materials from landfill disposal. This strategy always necessitates judgments about the relative benefits of available waste management options. This has come to be known as the waste management hierarchy, a structured ranking of priorities that directs decision-makers. The topic of waste management is intricate and has many different aspects. It is possible to quickly lose sight of the big picture. Waste management in Europe is particularly difficult since environmental protection must be accomplished without harming the continent's internal market. The MSn varies depending on a number of variables. Household waste is an indicator of economic status, population density, seasonality, housing standards, and the availability of waste minimization progrmes. The character of the business will have an impact on commercial waste.

7.3 THEORY AND PRACTICE

Landfill disposal is the only remaining management choice for leftover garbage. Everything else might be thought of as pre-treatment or waste avoidance in some way. To reduce the amount of material that can be considered to have been wasted, it is evident that all alternative management alternatives have a part to play. Indeed, some resource-conserving policymakers prefer to view landfill disposal as a long-term storage component of a broader Zero waste strategy rather than seeing it as the ultimate loss of resources. After waste is produced, the only alternatives left are to recover materials and/or energy sources before the useless remnants are dumped. The methods used rely on the types of waste being generated, the regional or local waste management systems that are available, and the market potential. The decision is largely influenced by the adopted waste management strategy. Although there were notable signs of convergence between the various factions in the late 1990s, disagreements about the relative advantages of management strategies for prevention vs disposal have been loud and heated. It is typically true that the primary goal of a waste management strategy is to lessen the likelihood that waste would affect the environment or the public's health, even when local conditions may change these objectives. Numerous sources make it abundantly evident that different nations choose to manage waste in various ways, or rather, exhibit various sets of preferences as alternatives to disposal. The most crucial method of managing solid wates is waste minimization, prevention, or avoidance because avoided waste doesn't require management and has no negative effects on the environment.

While general advice is beneficial in theory, it frequently makes little difference in any given situation beyond pointing out potential waste reduction strategies. It is first necessary to determine the composition of that waste and the causes that led to its formation in order to successfully reduce waste volumes. In a domestic setting, those factors may include or be determined by lifestyle, such as when both parents are employed full-time and must purchase more convenience meals, or when there is a young child in the family who wears disposable diapers. Smaller households are becoming more prevalent in some nations, which raises the likelihood that more garbage will be produced. For instance, a research conducted found that households with a single person can produce about 120 kilograms of food waste annually. In contrast, a two-person family produces only 85 kilograms of food waste per person each year. This amount drops to 50 kg per person per year in a three-person household. In a business setting, some waste may come from delivery guidelines established by a centralized supply chain or from decisions made on the types of machinery years earlier, necessitating a substantial investment in new equipment to modify. Home composting progmmes, in which homeowners compost their own kitchen and garden waste, can be regarded as waste minimization efforts, but all other waste handling processes, such as waste collection for recycling, while relieving pressure on disposal facilities, do not actually reduce the amount of waste produced. This reduction might be accomplished by altering the manufacturing processes, or even by altering the type or combination of materials employed.

It is simple to conceive of waste as a solid product that is left over after a process or activity, but it would be incorrect to focus only on minimizing the amount of material waste created without also taking into account other factors like energy or water waste. It is just as wasteful to throw away newspapers, cans, or empty wrappings as it is to use several liters of water needlessly, drive a car when walking or cycling would be more efficient, or use energy carelessly. For instance, traveling a distance to drop off certain newspapers, cans, glass, or plastic bottles at collection bins can be wasteful. The resources used in doing so, along with the additional resources required to transport the items for reprocessing, may outweigh the resources saved by keeping them out of the waste. The prospect of reusing or recycling a product can collide with the need to utilize le few resources as possible in its production. For instance, a refillable glass bottle needs to be tougher to be able to endure numerous trips between the consumer and the bottling facility, which requires more glass. The more travel required to move a container and its contents to their destination, the heavier it must bal The environmental effects of transportation are significant. As a result, the number of trips each bottle takes back must be carefully weighed against the added resource use associated with making it refillable. The resources wasted are larger than they would be if the bottle had been made for a single use if it is thrown away and not returned or only filled once or twicss Fewer materials are used in a single product, which makes recycling easier. For instance, the recent switch by automobile manufacturers to using only three or four varieties of plastic, instead of more than twenty previously, has made it easier to reclaim plastic from waste cars. Sometimes a product can perform its function with the least amount of resources by combining paper and plastic, or by covering glass with plastic, but because of the mixed materials, it is more difficult to recycle than, say, an item made entirely of paper (Ford & Despeisse, 2016).

7.4 PACKAGING

Despite the fact that the value of the wasted food is many times greater than the value of the packaging that is now avoided, extreme packaging reduction measures may have the unintended consequence of reducing the use of paper, metals, glass, and plastics at the expense of the food they would have protected. According to surveys, households in industrialized nations waste a lot of food simply because they buy too much. Demanding new clothing and furniture just to keep up with trends is wasteful because the current items still have functional life in them. Even though waste minimization is difficult for individuals and households to achieve, there are certain things that can be done. For instance, it is important to choose suitable quantities and sizes while making purchases of commodities. Two instances of needless waste formation include purchasing huge paint tins for a little decorating project or more food than can be eaten fresh. Since waste includes more than just the waste left out for pickup every week, there can be significant resource savings achieved by using less electricity, driving less, and other practices. By purchasing fewer items, items with a longer shelf life, and reusing items, empty tins and jars make wonderful storage, yogurt pots are perfect for seedlings, and magazines may be given to friends and neighbors after being read, individuals can limit the amount of waste they produce. Another significant contribution is made by mending damaged or worn-out clothing and/or equipment. Product durability can help reduce waste, and in the majority of cases, extending the usable life of a product like a vacuum cleaner to fifteen years from twelve can significantly aid in resource optimization. However, there are some circumstances in which extending an item's life could really be a bad idea for the environment as a whole. If you use your old washing machine frequently and it consumes twice as much water, detergent, and electricity as the newer model, keeping it in use may be worse for the environment in the long run than having it scrapped, having its metal recycled, and buying a new one. Another example is motor vehicles, which often use more fuel and emit more pollution than more recent models.

7.5 STRATEGIES FOR REDUCING WASTE

Strategies for reducing waste should begin with product design early in a product's life cycle. A product's ability to be repaired as well as its ability to be recycled at the end of its useful life can both be facilitated by minimizing the number of distinct components utilized in the product or by making it easier to disassemble.

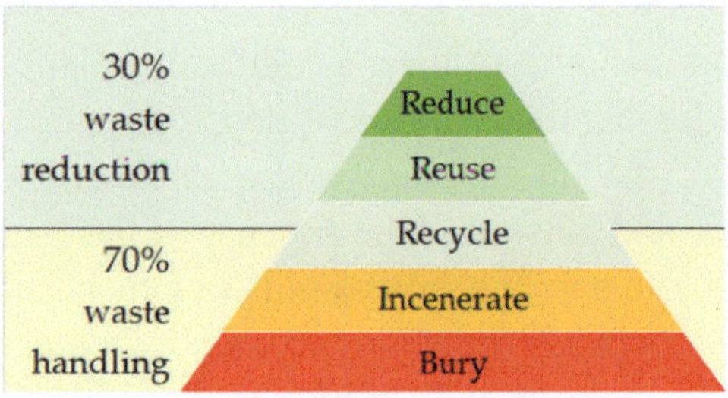

Figure 7.1: Waste Management targets for 2025 (Source: Ratnasari S, Mizuno K, Herdiansyah H, Simanjutak EGH. Enhancing Sustainability Development for Waste Management through National–Local Policy Dynamics. Sustainability. 2023; 15(8):6560. ***https://doi.org/10.3390/su15086560***)

Whether it is crucial to reduce the amount of raw resources needed to produce the items may need to be decided. It's possible that making things less hazardous to the environment while in use or minimizing the volume or toxicity of waste produced when the products have reached the end of their useful lives are given higher priority. Hiring or borrowing equipment that is not frequently needed offers a lot of potential for waste reduction. Examples of this include decoration tools and gardening tools like shredders. Another way to reduce waste is to use equipment in a shared or collective setting. For instance, a local neighborhood or community group may buy a commercial-sized washing machine together and operate their own laundry service, saving on both the amount of equipment needed and on electricity and water. This might be especially important in a neighborhood where many infants and young children use diapers. The minimization of waste can be addressed through a variety of policy tools. The recovery of materials, compost, or energy is largely influenced by the technique of collection; this, in turn, influences whether markets can be found. The point of interaction between generators and the waste management system is also collection. Since the type of collection influences the needs for sorting and certain collecting methods themselves involve some amount of sorting, collection is rarely independent of future sorting.

Blended collection of all garbage is likely the most practical solution from the householder's perspectve, but also restricts the alternatives for further treatment. The majority of treatment techniques require some sort of home separation of the waste into separate fractions before collection. In order to participate in kerbside collection progmmes, a householder must put recyclables in a container that is left outside the home on a designated day for collection. Commodities including bulky goods and garden garbage are transported by homeowners to a central collection site. These sites frequently feature recycling collection bins as well. Although kerbside collection is more specifically defined, collection of segregated fractions or mixed garbage is also possible. The different waste streams are collected separately, whether by the same collection truck or by a different one, in areas where household sorting has taken place. Separate collections are most frequently used in Europe and North America for dry recyclables, biowaste, and, in some nations, domestic hazardous waste. A collectfor of any residual garbage is required. Bulky garbage and garden waste can be handled separately or combined with the biowaste and other waste streams, respectively. Recyclables have been collected using a variety of kerbside collection techniques, depending on the amount of sorting required, and incudes boxes, bags, and wheeled bins. Recyclables are, in their most basic form, separated and kept all together in a bag, box, or wheeled container until collection. Similar to when mixed recyclables are collected from street containers, collection can be done with already-in-use collection vehicles, sometimes even with compaction. In order to recover commodity-grade materials for sale or to recover a mixed material fraction for further processing or conversion, solid waste is processed in materials recovery plants. The wastes may be transported to the facility as mixed solid wastes, as source-separated wastes, or in both forms, depending on the purpose and architecture of the Material Recovery Facility (MRF) (Daskalopoulos, Badr, & Probert, 1997). However, facilities typically treat either mixed waste or those that have been source-separated, not both. It is not always easy to tell the difference between mixed solid wastes and source-separated components. While source-separated materials are made up of specific parts intended for recycling, mixed solid waste is heterogeneous. The source-separated mixtures start to resemble mixed solid waste as more components are introduced to source-separated collection prrammes in an effort to recycle more resources.

7.6 MATERIAL RECOVERY FATYMRF

Both broad categories of MRFs can process materials mechanically, manually, or both ways. A town or area may need both source-separated and mixed waste MRFs in order to achieve high levels of waste diversion from landfill. Residential, commercial, institutional, industrial, or any combination of these mixed solid wastes can be processed by mixed waste MRFs. The transported wastes are anticipated to include high levels of contamination, and non-processable objects will need special management. Items that cannot be processed include bulky objects like rolled carpets, water heaters, and vehicle exhausts as well as materials like cable, fabrics, and magnetic tape. Due to the high levels of contamination and consequently poor yield of goods of marketable quality, it is reasonable to anticipate that the gross recovery rate for secondary materials of commodity grade will be low. The transformation of the raw feedstock into a useful end product is the MRF's main goal. The processing must be done in a way that ensures the finished goods meet a certain set of requirements. High-quality recovered materials are desired to ensure the continuation of end-product sales, particularly in times of tight markets. A facility for processing materials employs a range of machinery. In an MRF, conveyors are crucial pieces of machinery. Frequently used equipment includes shredders, crushers, magnetic separators, screens, and densifiers. Forklifts, trailers, and front-end loaders are all examples of rolling equipment used in MRFs. Bag opening procedures must be incorporated into the design of mixed waste MRFs in order to release their contents for segregation and potential recovery. Similar procedures are needed for source-separated MRFs that take bagged materials. Whether a mixed waste MRF or a source-separated MRF is used to recover recyclable materials, successful design and equipment selection must account for the particle size distribution and other pertinent properties of the materials to be processed, achieve the required level of purity and yield of end-product, and prepare the end-products to other specifications of the marketplace, such as, for example, bulk density. Compost is defined more scientifically as the end result of biological processes including bactfungusfungi, insects, and animals that take place in the presence of an appropriate supply of oxygen and botanical and putrescible waste. Complex organic materials are broken down biologically into carbon dioxide, water, and compost as a byproduct. The finished product is reasonably stable and can be used as a mulch, a growing media component, or to improve the soil.

When there is an abundance of oxygen, moisture, and heat, composting takes place. For the process to be successful, compost must be frequently aerated, either by turning the heap or by introducing air into the composting material. It is possible for composting waste to reach temperatures high enough to destroy pathogenic microbes, and pests under carefully managed settings. When the oxygen supply is limited, anaerobic fermentation takes place. Complex hydrocarbons are converted into reduced intermediate by-products, which release carbon dioxide and methane, both of which have potential as fuels. In landfills containing domestic waste, anaerobic decomposition, also known as fermentation or digestion, constantly takes place. The breakdown of waste is primarily mediated by three kinds of microbes: ba, fungusfungi, and actinomycetes. As the organic matter starts to rot, bacteria and fungi take over. The rate of metabolic activity of these microorganisms is so high that temperatures can reach 70°C or higher if there is enough air available. Only bacteria and actinomycetes that can withstand heat at this point can continue to break down the waste. The pace of decomposition gradually decreases as the substrate is used up, the compost cools, and the fungus and non-thermophilic bacteria start to function once again. Since all microorganisms require water to survive, microbial activity slows down if the moisture level of composting waste goes below 40%. However, if the moisture content is too high, water will accumulate in the air gaps inside the composting material, resulting in anaerobic conditions that can produce unpleasant odors. A significant amount of the organic waste generated by households can be separated out and composted.

7.7 PROCESSING

Pre-processing stage involves materials recovery, after which the waste is ground up or screened to isolate the organic component (Kumari & Samadder, 2022). The composting waste is piled up in long heaps that are about two meters high, and windrow composting uses mechanical turning to aerate it frequently. To more effectively aerate and remove heat from the windrows, blowers can be utilized to push air through them. The garbage is carefully heaped over a vented floor area or a perforated pipe in forced aeration systems so that air can circulate through. It's possible for the ventilation to be created via a vacuum, which draws air in through the waste. This enables any created odors to be contained and handled as necessary.

Figure 7.2: Composting of wastes (Source: Luskar L, Polanšek J, Hladnik A, Čeh B. On-Farm Composting of Hop Plant Green Waste—Chemical and Biological Value of Compost. Applied Sciences. 2022; 12(9):4190. https://doi.org/10.3390/app12094190)

Air can be blown through the composting waste in an outward direction; this approach is frequently chosen because it permits heat from the waste that is degrading most quickly near the center or base of the pile to be transferred to the cooler, outer areas. In comparison to windrow systems, forced aeration systems often have higher capital costs and lower operating expenses. As an alternative, a number of enclosed reactor systems that permit precise control over temperature, moisture, aeration, and mixing rates may be used. However, they also demand a significant capital investment. These more sophisticated technologies are often used with waste combinations that are more complicated. Complex organic molecules undergo a gradual breakdown into smaller components during decomposition. Crops may be poisoned by some of the intermediate breakdown products produced by this process. Before it can be used, the compost may need to go through a final maturation or curing step during which these plant toxins are further broken down. This could take several weeks or months.

Naturally, the amount of compost that a mixed municipal waste composting facility can create varies according to the garbage that is received, but it also depends on the composition of the waste and the standards set for the finished product. Due to the high moisture content of the majority of organic waste, composting has a drying impact. Dry weight also decreases significantly as a result of the conversion of organic carbon to gaseous carbon dioxide. The output of a composting facility is also impacted by the final steps in the processing. In order to remove inert pollutants like wood, glass, or plastic as well as uncomposted particles, a final sorting stage is typically required for high-grade compost. These last-stage screening procedures might not necessarily be required, for instance, if the compost will be utilized for landscaping, land reclamation, or landfill cover. It is difficult to market compost made from mixed household garbage. There have been numerous instances of market failure in Germany and other European nations. However, there has been such a high demand for the compost made in the Middle East that some potential customers have had to sign up for a two-year waiting list. Farmers nearly primarily use the compost made in Japan and the former USSR. Increasingly, governments are advocating the separate collection of putrescible household garbage. Separate collecting systems are required in the Netherlands as of 1994. Similar restrictions are currently in force in Austria, Germany, and Switzerland. Municipal authorities have been reluctant to invest in plants without a guaranteed market for the product because compost must compete with alternative soil conditioners.

Compost of poor quality, improperly stabilized, and including obvious impurities like glass and plastic has hurt markets. Even without a ready market, composting can still be a useful tool for lowering household waste's potential for pollution. By stabilizing biodegradable materials and reducing their volume, composting preserves landfill space and lowers the danger of leachate and landfill gas contamination. Before the waste is composted, it is challenging to sort out the primary sources of heavy metals, and work is being done to make the procedure better. Since the majority of the heavy metals are concentrated in the particles of household waste, proper screening can eliminate a significant amount of them. Complete eradication of these pollutants is rare, though.

7.8 POLLUTANT HANDNG APPROACHESS

Household dust, staples, and batteries are some of the hardest metal-rich pollutants to remove. It is significantly preferable to ensure that pollutants do not mix with compostable waste when producing high-quality composts, either via source separation for commercial composting or by composting at home. Organic waste can be composted at home by a single householder, which not only removes it from the waste stream but also gives them full discretion over what kind of waste goes on the compost heap, they are only limited to adding biodegradable materials. The local government can run a service in which the organic percentage is collected separately from other waste if householders are unable to compost their own waste. In this scenario, many of the issues are minimized, and the compost that is produced will be of a much higher consistency of quality. The creation of compost from MSW may be best done as part of an integrated waste management system if organic material is not separately collected. In Duisberg, Germany, such a facility has been in operation for thirty years. A magnet is used to separate ferrous metals from mixed household waste from a region before glass bottles, non-ferrous metals, and hard plastic objects are hand-selected along conveyor belts. The remaining waste is thereafter fed into a revolving reactor vessel where temperatures and moisture are meticulously regulated. After spending 36 hours in this reactor, the waste is filtered. The majority of the remaining compost is utilized for landscaping and land reclamation, with vineyards using the majority of it. Composting activity has significantly increased in Germany in recent years.

Composting facilities in France used to handle mixed household solid waste in the 1980s (D. Q. Zhang, Tan, & Gersberg, 2010). The outcome was a subpar product with sharp glass and plastic film, polluted with heavy metals. End markets for mixed waste compost shrank as farmers requested higher grade soil for application. A few municipalities implemented selective kerbside collection for biodegradable waste around the beginning of the 1990s. Composting facilities have been constructed across several Japanese cities to lessen the amount of garbage sent to incinerators. Around 29 composting units were in use in Japan by the middle of the 1990s. It's common for the waste stream to contain more organic material in less developed countries. Brazil currently has 74 composting plants, 19% of which are fully mechanized, and produces 90,000 tonnes of MSW per day with an organic content of up to 60%. Source separation for organic waste is used in some form by 30 municipalities in Brazil, significantly enhancing the quality of the finished product. With more compost being produced and used, there should be a rise in demand for compost prepared from source-separated materials. Since the topsoil is currently being lost at a pace of three tonnes per acre per year, it appears likely that compost will someday be in demand for repairing the structure of farm soils all over the world. It is being used for spreading on regions of saline land to restore fertility. The 1999 landfill directive of the European Union, which establishes strict guidelines for waste disposal, had a significant impact on the creationf these programmes.

According to a nationwide research by the Swiss Environment Agency, burning domestic garbage at home or in the garden is currently Switzerland's biggest source of dioxins. MSW combustion under controlled conditions produces an annual dioxin emission of 16g, compared to household waste burning, which produces 27–30g. Garbage treatment and disposal are the main contributing factors, and municipal waste incineration is the main source of dioxin production. There is debate about whether energy recovery and material recycling should be valued equally. Most individuals concur that energy recovery is a valuable alternative within an integrated approach to resource and waste management, but that materials recycling should typically be taken into account before energy recovery. Since many recyclable components, like glass, aluminum, and steel, are not combustible, removing them increases the efficiency of EFW plants. Operation of the plant is further aided by the removal of moist organic waste for composting. Before building a plant, waste streams and volumetric analysis should be done to minimize conflicts between material and energy recycling. By decomposing organic wastes in the absence of air at a landfill site, landfill gas is created. The usual composition of landfill gas is around 40% carbon dioxide and 55% methane, with traces of nitrogen, hydrogen, and water. A system of horizontal pipes and wells that are installed both before and during the filling of the site with garbage can be used to collect these gases. The problem of potentially explosive gas leaks from landfill sites led to the development of the advantageous use of landfill gas for energy. Utilizing methane for energy recovery has the added benefit of lowering the net potential for global warming because methane is a greenhouse gas with a higher global warming potential than its combustion byproduct, carbon dioxide.

7.9 ANAEROBIC DIGESTION

Anaerobic digestion can be used to break organic waste, and the methane gas that is produced can be captured. Agricultural wastes and sewage sludge have both been extensively processed using anaerobic digestion. Its usage to treat MSW, frequently in conjunction with sewage sludge, produces a fuel that may be cleaned and added to gas sources or utilized, like landfill gas, to directly fire burners and produce electricity. The ability to collect and use all of the gas produced by AD is a significant benefit over landfill gas, where collection efficiency is typically low. Additionally, digestate is a solid byproduct of AD that can be stored and utilized as fertilizer. Mass incineration, a traditional method of waste combustion, includes burning the garbage as it is received after bulky items have been removed. The waste is typically mixed to aid in combustion. Incineration facilities today are typically built to recover energy from the waste, as opposed to the past when they were primarily meant to process waste. It is not new to produce fuel from waste. It was first designed to transform MSW into a transportable, storage-capable fuel rather than burning it immediately. The following thermal conversion of combustible waste components is made possible by the generation of RDF. When producing RDF, the waste may go through several pre-processing processes, whereas mass burning necessitates little sorting or processing of the waste. RDF can be generated from mixed MSW using a number of screening steps along with the magnetic removal of ferrous and non-ferrous metals to create a coarse material. It could also be further processed into a densified, pelletized fuel for easier transport and storage. The two-step process of converting waste into coarse or pelletized RDF varies from mass burning in that the first processing stage can be carried out entirely independently from the second burning stage and may take place at a different location and time.

Fluidized bed combustion technique is based on a system where the fire bed is made up of inert particles like sand or ash rather than the waste being burned on a grate. The material of the bed behaves like a fluid when air is blowing through it. Fluidized bed combustors come in a variety of designs, such as circulating beds. Fluidized bed combustion systems have lower emissions, in part because of the process itself and in part because lime can be added to the bed. There are savings to be had since fluidized bed systems require less of an air pollution control system (APC), which can account for up to one-third of the cost of the plants. Mass burning facilities, on the other hand, don't require the front-end processing of the waste. Future MSW streams will see more Advanced Thermal Treatment (ATT) system applications, largely because these technologies are seen as more environmentally friendly and socially acceptable waste disposal options than traditional incineration with energy recovery. Waste can be transformed into molecular building blocks through the gasification and pyrolytic processes, creating new feedstock compounds for the petrochemndustry. Waste tyres tires and plastics can be recovered for material recycling and energy recrillo, & Cornacchia, 2000). In waste management, landfills are frequently seen as a last choice. This may be partially true, but the majority of solid wastes are currently treated actively through landfilling. Facilities that are appropriately planned may also result in additional advantages, such as material and energy recovery or land reclamation. Municipal solid wastes became an issue as urbanization grew. In addition to the inconvenience of having waste dumped in the streets, waste disposal also became important due to very serious health risks. Cities used to dump their waste in rivers and lakes for a while. Others simply dumped waste on vacant ground in open pits. Each landfill is classified according to the type of waste it receives using a common system. Landfills are available for municipal, inert, and hazardous garbage. In actuality, these definitions are not mutually exclusive. Other variations include mono-fills, where only one type of garbage is permitted, and co-disposal sites, where it is possible to mix municipal and hazardous wastes. Total containment entails a commitment to long-term monitoring and oversight. With hazardous wastes, this tactic is widely employed; for municipal solid garbage, it is less common. The use of a low-permeability liner beneath the wastes, as well as the collection, removal, and external treatment of liquid decomposition products, help prevent water leakage from the landfill. Leachate management effectiveness and the degree to which the containment barrier integrity is upheld determine the risk of leachate pollution.

7.10 LEACHATE HANDLING

Leachate levels are allowed to increase within the garbage, allowing for progressive migration through the liner into the ground, even if sumps for collection and removal of leachate are occasionally provided. It goes without saying that this strategy is not appropriate for every geographic and geological site, hence thorough risk and environmental impact assessments should be conducted before construction. The liner, its protective layer, a leachate drainage layer, and a top cover are often the main elements of containment. Sanitary landfilling calls for keeping wastes out of the environment until natural biological, chemical, and physical processes have rendered them harmless. The components of garbage that is landfilled might interact in a variety of complex ways. In general, organic waste makes up more than 50% of household waste. The behavior of garbage is influenced by compaction. This procedure starts with collecting, moves to the landfill location, and then continues inside the landfill. Water's impacts on garbage as it moves through the landfill, dissolving soluble components and moving unreacted debris, have a significant impact on how the waste behaves over time. Pollutants in the water are absorbed, which aids in the retention of items, at least before saturation. Adsorption plays a substantial role in a landfill as wastes bind to the surface of other materials. Above-ground landfills have increased recently. Landraising does have some visual advantages, but it also has environmental disadvantages. Leaks are simpler to spot and stop, and waste is kept farther away from any potential groundwater interaction. Leachate and landfill gas are the two main types of pollution that must be adequately controlled by a waste disposal facility. Some liquid is also collected along with landfill gas. Leachate is produced when moisture enters a landfill, whether it be by runoff, snowmelt, rain, or moisture in the waste itself. Heavy metals and organic compounds are the two leachate elements that matter most. They may be preserved in landfills, but they can be mobilized and released through biological processes. As byproducts, organic compounds are present. They are also present in the garbage as organic compounds. Leachate from landfills frequently contains heavy metals like mercury, chromium, nickel, lead, cadmium, copper, and zinc. Acidity and leachate flow rates are the key determinants of discharges. The non-regulated hazardous waste portions from homes and businesses contain a lot of heavy metals. There is no engineering design that can guarantee complete containment, and leachate may migrate and harm groundwater supplies. Policymakers typically demand the implementation of thorough environmental impact and risk analysis approaches as well as quality control systems. On-site treatment, discharge into sewer systems, or transport off-site for treatment elsewhere are the three main types of leachate management techniques.

Methane and carbon dioxide typically make up the majority of this gas at a variety of concentrations. Rapid composition changes in the gas mean that methane doesn't start to build up in substantial amounts for 3 to 12 months. Although initial degradation is aerobic and results in the production of carbon dioxide and water, hydrogen and hydrogen sulfide may also be formed at first. Later, the reaction turns anaerobic. There are several contaminants in landfill gas emissions that are dangerous to human health, including acrylonitrile, benzene, and carbon tetrachloride. Volatile organic chemicals found in landfill gas emissions contribute to urban fog. One of the main producers of methane, a powerful greenhouse gas, in the environment is landfills. Methane concentrations in the atmosphere have risen since the turn of the 19th century. It has been calculated that emissions from open dumps and landfills account for 6% of all methane emissions. Ventilation systems are employed to keep landfill gas from accumulating to unsafe levels. Full recovery plans typically involve collection, extraction, and transportation components and are added in stages as the landfill is built. Due to the fact that methane is less thick than air, some landfill gas will undoubtedly escape. Costs may be decreased by recovering and using methane. Nearly two-thirds of municipal waste dumps in Germany use landfill gas to generate electricity. When the landfill is finished, the top is sealed, the cells are capped with an impermeable membrane, and more than half a meter of earth is placed on top of the waste.

Many governments have made an effort to lessen their reliance on landfilling. A Landfill Tax was implemented by the British government, for instance, in 1996 (Morris & Read, 2001). In 2002, it was $12 per tonne for active waste and $2 per tonne for inert garbage. According to an investigation, the amount of active waste disposed of did not vary significantly, but the amount of inert garbage disposed of decreased by 20 to 30 Mt. Untreated items are often prohibited from landfills in many nations. According to a US poll, almost all states currently have restrictions in place, albeit not all of them are actually enforced. In California, restrictions apply to lead-acid batteries, adhesives, automotive products, latex paint, vehicles, lable metals, whole tyrestires, cleaners, pesticides, mercury, solvents, spent oil, and domestic batteries. The European Union Landfill Directive is a major piece of legislation that expands on the idea of the landfill ban. The fundamental requirement of this directive is the gradual restriction on municipal biodegradable waste from landfills, which will be reduced to 35 percent of levels in 1995 by 2020. Landfills currently handle 55% of the MSW produced in America. Municipal solid waste landfills are presently less prevalent in the US than they were in the 1980s, but their average size has grown. The proportion of MSW dumped has decreased consistently while recovery rates have risen and combustion has stayed constant. Some communities in Australia and Canada have promoted the idea of viewing landfills as long-term material storage facilities, ready to be used in the event that economic conditions change and turn waste into a valuable resource. Maybe this isn't any more sustainable than the alternatives if such a facility requires long-term environmental monitoring.

No matter how hard we work to reduce waste, enhance reuse and recycling, compost, and recover energy, there will always be some residual material that society cannot use in any other way. To encourage resource usage that is more efficient and the simultaneous economic, social, and environmental goals of sustainable development, new, integrated systems are needed. Waste can be handled in many ways to reduce threats to the environment and to public health. Waste management was done in a fragmented, often improvised manner for a long time. A strategy in which decisions on waste policies and practices take into account waste collection, treatment, and disposal methods, environmental advantages, economic optimization, and social acceptability is known as integrated waste management. The goal of integrated waste management (IWM) systems is to achieve environmental benefits, economic optimization, and societal acceptability by combining waste streams, waste collection, treatment, and disposal methods. IWM is a concept with various local applications that depends on a range of factors, including the composition of the waste stream, infrastructure, markets for recyclables, budgets, regional laws, and the availability of land. IWM looks for the finest waste management choices, with a focus on assessing all feasible tactics to produce more sustainable systems. The potential facets of this integration are many. It is important to combine upstream collection and handling with downstream treatment, processing, and disposal. Municipal solid waste management can be combined with commercial and industrial waste sources. For the system to be sustainable, recycling and composting collections must be connected with product markets. The creation of new waste management facilities is made possible by a higher level of integration found at the policy level. Collaboration between the public, industry, and planning authorities is required here. Everywhere in the world, the IWM concept has different specifics. IWM, for instance, includes efforts to promote markets for recyclables as well as public edution and outreach programmes in California. IWM in South Africa strives to integrate and optimize waste management in order to maximize effectiveness, minimize negative effects on the environment and financial expenses, and raise the quality of life.

7.11 INTEGRATED WASTE MANAGEMENT

(IWM) refers to the entire approach to thinking about how to best manage wastes and does not explain or prescribe the actual strategies used to deal with a specific waste stream. A town or area need not use every waste management technology; an integrated approach simply means that they are all being examined. Because waste is unpredictable, no one solution has ever been able to handle an entire waste stream. There are boundaries beyond which each alternative becomes technically impossible, environmentally harmful, economically inefficient, and socially unacceptable. Policies can offer a framework where more r less waste may be channelled in specific directions. While the organic component of garbage rarely reaches 70%, the highest limits for recycling are rarely more than 60%. More frequently than not, less than 60% of municipal solid waste is combustible. Any waste management system operations are inextricably linked. For instance, the kind of sorting system implemented in a region would influence the capacity to recover materials or create compost that can be sold. Many scessful recycling programmes produce residual garbage that is better suited for energy recovery. Reclaimed garbage from landfills and landfill gas can both be obtained. In general, the waste management costs are thought to be the responsibility of the waste-generating owner at the disposal site. The earliest manufacturers, importers, or sellers are frequently the ones who bear the costs, according to policymakers. This was the situation with European legislation on end-of-life automobiles and waste electrical and electronic equipment. The European Commission concluded that the final user should not incur any disposal charges that would deter proper waste management when drafting laws on various waste streams. Shared responsibility places the burden of waste management costs on the product owner at every step of the product's life cycle and offers each actor in the life cycle a clear financial incentive to limit the quantity of waste they produce.

Municipal solid waste management and collection services are provided in large part by local governments. In these situations, homeowners directly cover the cost of the services. The typical means of funding the current waste management systems are municipal taxes and fees. These, however, are frequently fixed, which again offers no incentive for homeowners to reduce waste. Local governments are increasingly using variable rate billing, which involves taxing homeowners in proportion to the amount of waste they generate. Variable rate billing does encourage more people to participate in recying and composting programmes or actively look for products that produce less waste. Early in 2002, it became evident that the Irish government is enthusiastally pursuing this programme and other tools of the market. Within three years, households in Ireland paid based on the weight of the garbage they placed out for collection, per a new policy designed to increase incentives for reuse and recycling. Variable rate billing structures can increase recycling rates while decreasing the amount of household waste produced and the real per capita cost of the service. The authorities in the Belgian province of Hainaut have to use variable rate billing to raise additional funds due to rising landfill expenses and high waste collection taxes. Unit pricing reduced the amount of garbage sent to the landfill by 40%. In some places, peopleo disposed of more waste to in landfills than allowed had to pay a fine.

A rising body of research suggests that IWM is being implemented at the local level, leading to higher levels of recycling and energy recovery and lowering the reliance on landfills currently in use. The specifics of integrated waste management systems are still not fully agreed upon, but this is to be expected because IWM is intended for local application, hence systems will vary. Also uncertain is whether the majority of legislative bodies will ever support a procedure that, by its very nature, tends to promote a more flexible approach to waste management. The adoption of rules and norms may be more challenging when there is flexibility. There is little debate, however, that taking a dogmatic strategy that forces the pursuit of aims that may not be justifiable is more likely to produce social, economic, and environmental benefits than flexible, responsive, and planned systems. In the 1990s, the dogma and argument about the relative importance of recycling and energy recovery diminished significantly. There is now broad consensus among proponents of energy recovery that materials from waste should first be recycled whenever possible before turning to energy recovery. On the other hand, many recyclers are aware thamaterials recovery programmes can be complemented by incinerators that are correctly sized and situated. IWM is both a way of thinking for decision-makers and other stakeholders and a useful strategy for standardizing problem-solving. Systems for managing waste have developed from simple precautions meant to safeguard human health to more complex systems focused on preserving the environment. The recent push for sustainable development has made it necessary to look at waste management from the environmental, economic, and social viewpoints, which has resulted in a larger use of IWM. Many consumers are irritated by the amount of packaging seen on grocery goods, despite the fact that they feel they cannot escape it. Some people's annoyance stems from environmental concerns, while others are offended because they think packaging misleads people about the things within. Plastic grocery bags are rarely reused, and this is often not a problem in homes. While some parents donated packaging to their child's school, the majority of people claimed to donate their used clothing to friends, relatives, or stores. Londoners don't generally believe that producing less garbage is their responsibility or something they can affect. Most individuals consider package reduction to be the responsibility of manufacturers and merchants, who should, if necessary, be persuaded through legislation. Even though there will always be some households that won't recycle and some that won't listen to instructions, the vast majority of people think they can do more with the appropriate kind of support.

If recycling goals are to be attained, increasing recycling convenience will be essential. This includes adding more kerbside recycling options and close-by recycling centers. Delivering services that people dse will also need involving to involve the customers in discussions about what offering is best for them. However, providing more services to more people won't be sufficient on its own to compel the necessary level of recycling and waste minimization. Neither will merely inform individuals of the existence of services. It will be necessary to educate people that recycling and minimization are normal behaviors, not only good behaviors. They must also be convinced that everyone is trying, so that participation is equitable, and that local governments are willing to do their bit. This means that in addition to offering exceptional services, you must also be perceived as doing so, either directly by providing feedback or inadvertently by the manner in which services are provided. Using examples from various instances where people and businesses have been convinced to adopt more sustainable behavior, it is suggested that more compelling strategies involve building progressively on tiny successes rather than aiming to make a sudden, huge leap.

References

1. Abd El-Salam, M. M. (2010). Hospital waste management in El-Beheira Governorate, Egypt. Journal of Environmental Management, 91(3), 618–629. https://doi.org/10.1016/j.jenvman.2009.08.012
2. Abd'Razack, N. T. A., Medayese, S. O., Shaibu, S. I., & Adeleye, B. M. (2017). Habits and benefits of recycling solid waste among households in Kaduna, North West Nigeria. Sustainable Cities and Society, 28, 297–306. https://doi.org/10.1016/j.scs.2016.10.004
3. Allesch, A., & Brunner, P. H. (2015). Material Flow Analysis as a Decision Support Tool for Waste Management: A Literature Review. Journal of Industrial Ecology, 19(5), 753–764. https://doi.org/10.1111/jiec.12354
4. Al-Maaded, M., Madi, N. K., Kahraman, R., Hodzic, A., & Ozerkan, N. G. (2012). An Overview of Solid Waste Management and Plastic Recycling in Qatar. Journal of Polymers and the Environment, 20(1), 186–194. https://doi.org/10.1007/s10924-011-0332-2
5. Altaf, M. A., & Deshazo, J. R. (1996). Household demand for improved solid waste management: A case study of Gujranwala, Pakistan. World Development, 24(5), 857–868. https://doi.org/10.1016/0305-750X(96)00006-X
6. Andreola, F., Barbieri, L., Corradi, A., & Lancellotti, I. (2007). CRT glass state of the art: A case study: Recycling in ceramic glazes. Journal of the European Ceramic Society, 27(2), 1623–1629. https://

doi.org/10.1016/j.jeurceramsoc.2006.05.009

7. Ardakani, M. N., & Badalians Gholikandi, G. (2020). Microbial fuel cells (MFCs) in integration with anaerobic treatment processes (AnTPs) and membrane bioreactors (MBRs) for simultaneous efficient wastewater/sludge treatment and energy recovery -A state-of-the-art review. Biomass and Bioenergy, 141, 105726. https://doi.org/10.1016/j.biombioe.2020.105726
8. Arena, U. (2012). Process and technological aspects of municipal solid waste gasification. A review. Waste Management, 32(4), 625–639. https://doi.org/10.1016/j.wasman.2011.09.025
9. Aryampa, S., Maheshwari, B., Sabiiti, E., Bateganya, N. L., & Bukenya, B. (2019). Status of Waste Management in the East African Cities: Understanding the Drivers of Waste Generation, Collection and Disposal and Their Impacts on Kampala City's Sustainability. Sustainability, 11(19), 5523. https://doi.org/10.3390/su11195523
10. Azam, M., Jahromy, S. S., Raza, W., Raza, N., Lee, S. S., Kim, K.-H., & Winter, F. (2020). Status, characterization, and potential utilization of municipal solid waste as renewable energy source: Lahore case study in Pakistan. Environment International, 134, 105291. https://doi.org/10.1016/j.envint.2019.105291
11. Batchelor, B. (2006). Overview of waste stabilization with cement. Waste Management, 26(7), 689–698. https://doi.org/10.1016/j.wasman.2006.01.020
12. Beierle, T. C. (2002). Democracy in Practice: Public Participation in Environmental Decisions. New York: Routledge. https://doi.org/10.4324/9781936331017
13. Bergman, K., & Appelgren, L.-E. (1983). Application and Results of Whole-Body Autoradiography in Distribution Studies of Organic Solvents. CRC Critical Reviews in Toxicology, 12(1), 59–118. https://doi.org/10.3109/10408448309029318
14. Blake, D. M. (1988). Solar thermal technology for the destruction of CFC waste. International Journal of Refrigeration, 11(4), 239–242. https://doi.org/10.1016/0140-7007(88)90080-1
15. Broun, R., & Sattler, M. (2016). A comparison of greenhouse gas emissions and potential electricity recovery from conventional and bioreactor landfills. Journal of Cleaner Production, 112, 2664–2673. https://doi.org/10.1016/j.jclepro.2015.10.010

16. Burian, S. J., Nix, S. J., Pitt, R. E., & Durrans, S. R. (2000). Urban Wastewater Management in the United States: Past, Present, and Future. Journal of Urban Technology, 7(3), 33–62. https://doi.org/10.1080/713684134
17. Busch, J. (1987). Second time around: A look at bottle reuse. Historical Archaeology, 21(1), 67–80. https://doi.org/10.1007/BF03374080
18. Chatterjee, A., & Sui, T. (2019). Alternative fuels – Effects on clinker process and properties. Cement and Concrete Research, 123, 105777. https://doi.org/10.1016/j.cemconres.2019.105777
19. Chen, Y., Li, J., Liu, L., & Zhao, N. (2012). Polybrominated diphenyl ethers fate in China: A review with an emphasis on environmental contamination levels, human exposure and regulation. Journal of Environmental Management, 113, 22–30. https://doi.org/10.1016/j.jenvman.2012.08.003
20. Cheng, K., Hao, W., Wang, Y., Yi, P., Zhang, J., & Ji, W. (2020). Understanding the emission pattern and source contribution of hazardous air pollutants from open burning of municipal solid waste in China. Environmental Pollution, 263, 114417. https://doi.org/10.1016/j.envpol.2020.114417
21. Cohen, B. (2006). Urbanization in developing countries: Current trends, future projections, and key challenges for sustainability. Technology in Society, 28(1), 63–80. https://doi.org/10.1016/j.techsoc.2005.10.005
22. Curtis, L., Rea, W., Smith-Willis, P., Fenyves, E., & Pan, Y. (2006). Adverse health effects of outdoor air pollutants. Environment International, 32(6), 815–830. https://doi.org/10.1016/j.envint.2006.03.012
23. Darby, L., & Obara, L. (2005). Household recycling behaviour and attitudes towards the disposal of small electrical and electronic equipment. Resources, Conservation and Recycling, 44(1), 17–35. https://doi.org/10.1016/j.resconrec.2004.09.002
24. Daskalopoulos, E., Badr, O., & Probert, S. D. (1997). Economic and Environmental Evaluations of Waste Treatment and Disposal Technologies for Municipal Solid Waste. Applied Energy, 58(4), 209–255. https://doi.org/10.1016/S0306-2619(97)00053-6
25. Dehghani, M. H., Omrani, G. A., & Karri, R. R. (2021). Chapter 11—Solid Waste—Sources, Toxicity, and Their Consequences to Human Health. In R. R. Karri, G. Ravindran, & M. H. Dehghani (Eds.), Soft

Computing Techniques in Solid Waste and Wastewater Management (pp. 205–213). Elsevier. https://doi.org/10.1016/B978-0-12-824463-0.00013-6

26. Dimmen, M. V., Steindal, E. H., & Basu, A. R. (2023). Overview of Use and Release Sources of the Seven New POPs. In P. Chakraborty, L. Nizzetto, G. Bharat, E. Steindal, & S. Sinha (Eds.), Managing Persistent Organic Pollutants in India: Case Studies on Vapi and Surat, Gujarat (pp. 153–180). Cham: Springer Nature Switzerland. https://doi.org/10.1007/978-3-031-31311-0_8
27. Ferraris, A., Mazzoleni, A., Devalle, A., & Couturier, J. (2018). Big data analytics capabilities and knowledge management: Impact on firm performance. Management Decision, 57(8), 1923–1936. https://doi.org/10.1108/MD-07-2018-0825
28. Ferronato, N., Moresco, L., Guisbert Lizarazu, G. E., Gorritty Portillo, M. A., Conti, F., & Torretta, V. (2023). Comparison of environmental impacts related to municipal solid waste and construction and demolition waste management and recycling in a Latin American developing city. Environmental Science and Pollution Research, 30(4), 8548–8562. https://doi.org/10.1007/s11356-021-16968-8
29. Ferronato, N., & Torretta, V. (2019). Waste Mismanagement in Developing Countries: A Review of Global Issues. International Journal of Environmental Research and Public Health, 16(6), 1060. https://doi.org/10.3390/ijerph16061060
30. Ford, S., & Despeisse, M. (2016). Additive manufacturing and sustainability: An exploratory study of the advantages and challenges. Journal of Cleaner Production, 137, 1573–1587. https://doi.org/10.1016/j.jclepro.2016.04.150
31. Francou, C., Linères, M., Derenne, S., Villio-Poitrenaud, M. L., & Houot, S. (2008). Influence of green waste, biowaste and paper–cardboard initial ratios on organic matter transformations during composting. Bioresource Technology, 99(18), 8926–8934. https://doi.org/10.1016/j.biortech.2008.04.071
32. Freeman, J., & Vermeule, A. (2007). Massachusetts v EPA: From Politics to Expertise. The Supreme Court Review, 2007, 51–110. https://doi.org/10.1086/655170
33. Friedrich, E., & Trois, C. (2013). GHG emission factors developed for the collection, transport and landfilling of municipal waste in South African municipalities. Waste Management, 33(4), 1013–1026. https://

doi.org/10.1016/j.wasman.2012.12.011

34. Gentil, E. C., Damgaard, A., Hauschild, M., Finnveden, G., Eriksson, O., Thorneloe, S., … Christensen, T. H. (2010). Models for waste life cycle assessment: Review of technical assumptions. Waste Management, 30(12), 2636–2648. https://doi.org/10.1016/j.wasman.2010.06.004
35. Getahun, T., Mengistie, E., Haddis, A., Wasie, F., Alemayehu, E., Dadi, D., … Van der Bruggen, B. (2012). Municipal solid waste generation in growing urban areas in Africa: Current practices and relation to socioeconomic factors in Jimma, Ethiopia. Environmental Monitoring and Assessment, 184(10), 6337–6345. https://doi.org/10.1007/s10661-011-2423-x
36. Gibbs, H. K., Ruesch, A. S., Achard, F., Clayton, M. K., Holmgren, P., Ramankutty, N., & Foley, J. A. (2010). Tropical forests were the primary sources of new agricultural land in the 1980s and 1990s. Proceedings of the National Academy of Sciences, 107(38), 16732–16737. https://doi.org/10.1073/pnas.0910275107
37. Goldman, S. L., & Nagel, R. N. (1993). Management, technology and agility: The emergence of a new era in manufacturing. International Journal of Technology Management, 8(1–2), 18–38. https://doi.org/10.1504/IJTM.1993.025758
38. Greco, G., Allegrini, M., Del Lungo, C., Gori Savellini, P., & Gabellini, L. (2015). Drivers of solid waste collection costs. Empirical evidence from Italy. Journal of Cleaner Production, 106, 364–371. https://doi.org/10.1016/j.jclepro.2014.07.011
39. Guerrero, L. A., Maas, G., & Hogland, W. (2013). Solid waste management challenges for cities in developing countries. Waste Management, 33(1), 220–232. https://doi.org/10.1016/j.wasman.2012.09.008
40. Guo, M., Song, W., & Buhain, J. (2015). Bioenergy and biofuels: History, status, and perspective. Renewable and Sustainable Energy Reviews, 42, 712–725. https://doi.org/10.1016/j.rser.2014.10.013
41. Hafid, H. S., Rahman, N. A. A., Shah, U. K. M., Baharuddin, A. S., & Ariff, A. B. (2017). Feasibility of using kitchen waste as future substrate for bioethanol production: A review. Renewable and Sustainable Energy Reviews, 74, 671–686. https://doi.org/10.1016/j.rser.2017.02.071
42. Hahladakis, J. N., Velis, C. A., Weber, R., Iacovidou, E., & Purnell, P. (2018). An overview of chemical additives present in plastics:

Migration, release, fate and environmental impact during their use, disposal and recycling. Journal of Hazardous Materials, 344, 179–199. https://doi.org/10.1016/j.jhazmat.2017.10.014

43. Haider, T. P., Völker, C., Kramm, J., Landfester, K., & Wurm, F. R. (2019). Plastics of the Future? The Impact of Biodegradable Polymers on the Environment and on Society. Angewandte Chemie International Edition, 58(1), 50–62. https://doi.org/10.1002/anie.201805766
44. Hansen, A. M., Kraus, T. E. C., Pellerin, B. A., Fleck, J. A., Downing, B. D., & Bergamaschi, B. A. (2016). Optical properties of dissolved organic matter (DOM): Effects of biological and photolytic degradation. Limnology and Oceanography, 61(3), 1015–1032. https://doi.org/10.1002/lno.10270
45. Harvey, S. (1983). Title iii of the marine protection, research and sanctuaries act: Issues in program implementation. Coastal Zone Management Journal, 11(3), 169–197. https://doi.org/10.1080/08920758309361944
46. Hayden, B., & Cannon, A. (1983). Where the garbage goes: Refuse disposal in the Maya Highlands. Journal of Anthropological Archaeology, 2(2), 117–163. https://doi.org/10.1016/0278-4165(83)90010-7
47. Hazra, T., & Goel, S. (2009). Solid waste management in Kolkata, India: Practices and challenges. Waste Management, 29(1), 470–478. https://doi.org/10.1016/j.wasman.2008.01.023
48. Hermansson, H. (2007). The Ethics of NIMBY Conflicts. Ethical Theory and Moral Practice, 10(1), 23–34. https://doi.org/10.1007/s10677-006-9038-2
49. Huntzinger, D. N., & Eatmon, T. D. (2009). A life-cycle assessment of Portland cement manufacturing: Comparing the traditional process with alternative technologies. Journal of Cleaner Production, 17(7), 668–675. https://doi.org/10.1016/j.jclepro.2008.04.007
50. Jagaba, A. H., Kutty, S. R. M., Lawal, I. M., Abubakar, S., Hassan, I., Zubairu, I., … Noor, A. (2021). Sequencing batch reactor technology for landfill leachate treatment: A state-of-the-art review. Journal of Environmental Management, 282, 111946. https://doi.org/10.1016/j.jenvman.2021.111946
51. Jang, Y.-C. (2010). Waste electrical and electronic equipment (WEEE) management in Korea: Generation, collection, and recycling systems. Journal of Material Cycles and Waste Management, 12(4), 283–294. https://doi.org/10.1007/s10163-010-0298-5

52. Jayal, A. D., Badurdeen, F., Dillon, O. W., & Jawahir, I. S. (2010). Sustainable manufacturing: Modeling and optimization challenges at the product, process and system levels. CIRP Journal of Manufacturing Science and Technology, 2(3), 144–152. https://doi.org/10.1016/j.cirpj.2010.03.006
53. Jha, A. K., Sharma, C., Singh, N., Ramesh, R., Purvaja, R., & Gupta, P. K. (2008). Greenhouse gas emissions from municipal solid waste management in Indian mega-cities: A case study of Chennai landfill sites. Chemosphere, 71(4), 750–758. https://doi.org/10.1016/j.chemosphere.2007.10.024
54. Johansson, J., & Luttropp, C. (2009). Material hygiene: Improving recycling of WEEE demonstrated on dishwashers. Journal of Cleaner Production, 17(1), 26–35. https://doi.org/10.1016/j.jclepro.2008.02.010
55. Kale, G., Kijchavengkul, T., Auras, R., Rubino, M., Selke, S. E., & Singh, S. P. (2007). Compostability of Bioplastic Packaging Materials: An Overview. Macromolecular Bioscience, 7(3), 255–277. https://doi.org/10.1002/mabi.200600168
56. Kamran, A., Chaudhry, M. N., & Batool, S. A. (2015). Effects of socio-economic status and seasonal variation on municipal solid waste composition: A baseline study for future planning and development. Environmental Sciences Europe, 27(1), 16. https://doi.org/10.1186/s12302-015-0050-9
57. Karak, T., Bhagat, R. M., & Bhattacharyya, P. (2012). Municipal Solid Waste Generation, Composition, and Management: The World Scenario. Critical Reviews in Environmental Science and Technology, 42(15), 1509–1630. https://doi.org/10.1080/10643389.2011.569871
58. Karekezi, S. (2002). Poverty and energy in Africa—A brief review. Energy Policy, 30(11), 915–919. https://doi.org/10.1016/S0301-4215(02)00047-2
59. Karouach, F., & El Bari, H. (2023). Waste Generation, Characteristics, and Collection in Developing Countries. In H. El Bari & C. Trois (Eds.), Waste Management in Developing Countries (pp. 1–21). Cham: Springer International Publishing. https://doi.org/10.1007/978-3-031-28001-6_1
60. Kaseva, M. E., & Mbuligwe, S. E. (2005). Appraisal of solid waste collection following private sector involvement in Dar es Salaam city, Tanzania. Habitat International, 29(2), 353–366. https://doi.org/10.1016/j.habitatint.2003.12.003

61. Kassim, S. M., & Ali, M. (2006). Solid waste collection by the private sector: Households' perspective—Findings from a study in Dar es Salaam city, Tanzania. Habitat International, 30(4), 769–780. https://doi.org/10.1016/j.habitatint.2005.09.003
62. Kohrell, C. R. R., Leander J. Schwartz, Robert B. Wenger, Mary G. (2020). Waste Management and Resource Recovery. Boca Raton: CRC Press. https://doi.org/10.1201/9780203734278
63. Koul, B., Yakoob, M., & Shah, M. P. (2022). Agricultural waste management strategies for environmental sustainability. Environmental Research, 206, 112285. https://doi.org/10.1016/j.envres.2021.112285
64. Kublik, H., & Vidgren, M. T. (1998). Nasal delivery systems and their effect on deposition and absorption. Advanced Drug Delivery Reviews, 29(1), 157–177. https://doi.org/10.1016/S0169-409X(97)00067-7
65. Kumar, S., Bhattacharyya, J. K., Vaidya, A. N., Chakrabarti, T., Devotta, S., & Akolkar, A. B. (2009). Assessment of the status of municipal solid waste management in metro cities, state capitals, class I cities, and class II towns in India: An insight. Waste Management, 29(2), 883–895. https://doi.org/10.1016/j.wasman.2008.04.011
66. Kumar, S., Dhar, H., Nair, V. V., Bhattacharyya, J. K., Vaidya, A. N., & Akolkar, A. B. (2016). Characterization of municipal solid waste in high-altitude sub-tropical regions. Environmental Technology, 37(20), 2627–2637. https://doi.org/10.1080/09593330.2016.1158322
67. Kumari, R., & Samadder, S. R. (2022). A critical review of the pre-processing and metals recovery methods from e-wastes. Journal of Environmental Management, 320, 115887. https://doi.org/10.1016/j.jenvman.2022.115887
68. Lampe, M., & Gazda, G. M. (1995). Green marketing in Europe and the United States: An evolving business and society interface. International Business Review, 4(3), 295–312. https://doi.org/10.1016/0969-5931(95)00011-N
69. Lara-Valencia, F., Harlow, S. D., Lemos, M. C., & Denman, C. A. (2009). Equity dimensions of hazardous waste generation in rapidly industrialising cities along the United States–Mexico border. Journal of Environmental Planning and Management, 52(2), 195–216. https://doi.org/10.1080/09640560802666545
70. Lee, C. K. M., Ng, K. K. H., Kwong, C. K., & Tay, S. T. (2019). A system dynamics model for evaluating food waste management in

Hong Kong, China. Journal of Material Cycles and Waste Management, 21(3), 433–456. https://doi.org/10.1007/s10163-018-0804-8

71. Liang, X., Goh, H. H., Kurniawan, T. A., Zhang, D., Dai, W., Liu, H., … Goh, K. C. (2022). Utilizing landfill gas (LFG) to electrify digital data centers in China for accelerating energy transition in Industry 4.0 era. Journal of Cleaner Production, 369, 133297. https://doi.org/10.1016/j.jclepro.2022.133297
72. Malinauskaite, J., Jouhara, H., Czajczyńska, D., Stanchev, P., Katsou, E., Rostkowski, P., … Spencer, N. (2017). Municipal solid waste management and waste-to-energy in the context of a circular economy and energy recycling in Europe. Energy, 141, 2013–2044. https://doi.org/10.1016/j.energy.2017.11.128
73. Marshall, R. E., & Farahbakhsh, K. (2013). Systems approaches to integrated solid waste management in developing countries. Waste Management, 33(4), 988–1003. https://doi.org/10.1016/j.wasman.2012.12.023
74. Massoud, M. A., Tarhini, A., & Nasr, J. A. (2009). Decentralized approaches to wastewater treatment and management: Applicability in developing countries. Journal of Environmental Management, 90(1), 652–659. https://doi.org/10.1016/j.jenvman.2008.07.001
75. Masud, M. H., Akram, W., Ahmed, A., Ananno, A. A., Mourshed, M., Hasan, M., & Joardder, M. U. H. (2019). Towards the effective E-waste management in Bangladesh: A review. Environmental Science and Pollution Research, 26(2), 1250–1276. https://doi.org/10.1007/s11356-018-3626-2
76. Memon, M. A. (2010). Integrated solid waste management based on the 3R approach. Journal of Material Cycles and Waste Management, 12(1), 30–40. https://doi.org/10.1007/s10163-009-0274-0
77. Mor, S., & Ravindra, K. (2023). Municipal solid waste landfills in lower- and middle-income countries: Environmental impacts, challenges and sustainable management practices. Process Safety and Environmental Protection, 174, 510–530. https://doi.org/10.1016/j.psep.2023.04.014
78. Morris, J. R., & Read, A. D. (2001). The UK landfill tax and the landfill tax credit scheme: Operational weaknesses. Resources, Conservation and Recycling, 32(3), 375–387. https://doi.org/10.1016/S0921-3449(01)00072-6
79. Mrkajić, V., Stanisavljevic, N., Wang, X., Tomas, L., & Haro, P. (2018).

Efficiency of packaging waste management in a European Union candidate country. Resources, Conservation and Recycling, 136, 130–141. https://doi.org/10.1016/j.resconrec.2018.04.008

80. Mudali, U. K., Patil, M., Saravanabhavan, R., & Saraswat, V. K. (2021). Review on E-Waste Recycling: Part I—A Prospective Urban Mining Opportunity and Challenges. Transactions of the Indian National Academy of Engineering, 6(3), 547–568. https://doi.org/10.1007/s41403-021-00216-z
81. Mupambwa, H. A., & Mnkeni, P. N. S. (2018). Optimizing the vermicomposting of organic wastes amended with inorganic materials for production of nutrient-rich organic fertilizers: A review. Environmental Science and Pollution Research, 25(11), 10577–10595. https://doi.org/10.1007/s11356-018-1328-4
82. Nakamaru, M., Iwasa, Y., & Nakanishi, J. (2003). Extinction risk to bird populations caused by DDT exposure. Chemosphere, 53(4), 377–387. https://doi.org/10.1016/S0045-6535(03)00010-9
83. Nath, B., & Marinova, L. (2000). Sustainable Solid Waste Management In The Former Yougoslav Republic Of Macedonia: A Nato Initiative. In B. Nath (Ed.), Sustainable Solid Waste Management in the Southern Black Sea Region (pp. 81–102). Dordrecht: Springer Netherlands. https://doi.org/10.1007/978-94-010-0940-9_5
84. Nehrenheim, E., Waara, S., & Johansson Westholm, L. (2008). Metal retention on pine bark and blast furnace slag – On-site experiment for treatment of low strength landfill leachate. Bioresource Technology, 99(5), 998–1005. https://doi.org/10.1016/j.biortech.2007.03.006
85. Nguyen-Trong, K., Nguyen-Thi-Ngoc, A., Nguyen-Ngoc, D., & Dinh-Thi-Hai, V. (2017). Optimization of municipal solid waste transportation by integrating GIS analysis, equation-based, and agent-based model. Waste Management, 59, 14–22. https://doi.org/10.1016/j.wasman.2016.10.048
86. Oenema, O., Wrage, N., Velthof, G. L., van Groenigen, J. W., Dolfing, J., & Kuikman, P. J. (2005). Trends in Global Nitrous Oxide Emissions from Animal Production Systems. Nutrient Cycling in Agroecosystems, 72(1), 51–65. https://doi.org/10.1007/s10705-004-7354-2
87. Ongondo, F. O., Williams, I. D., & Cherrett, T. J. (2011). How are WEEE doing? A global review of the management of electrical and electronic wastes. Waste Management, 31(4), 714–730. https://doi.org/10.1016/j.wasman.2010.10.023

88. Padmanabhan, K. K., & Barik, D. (2019). Chapter 8—Health Hazards of Medical Waste and its Disposal. In D. Barik (Ed.), Energy from Toxic Organic Waste for Heat and Power Generation (pp. 99–118). Woodhead Publishing. https://doi.org/10.1016/B978-0-08-102528-4.00008-0

89. Pagdee, A., Kim, Y., & Daugherty, P. J. (2006). What Makes Community Forest Management Successful: A Meta-Study From Community Forests Throughout the World. Society & Natural Resources, 19(1), 33–52. https://doi.org/10.1080/08941920500323260

90. Pattnaik, S., & Reddy, M. V. (2010). Assessment of Municipal Solid Waste management in Puducherry (Pondicherry), India. Resources, Conservation and Recycling, 54(8), 512–520. https://doi.org/10.1016/j.resconrec.2009.10.008

91. Powell, J. T., Townsend, T. G., & Zimmerman, J. B. (2016). Estimates of solid waste disposal rates and reduction targets for landfill gas emissions. Nature Climate Change, 6(2), 162–165. https://doi.org/10.1038/nclimate2804

92. Pujara, Y., Pathak, P., Sharma, A., & Govani, J. (2019). Review on Indian Municipal Solid Waste Management practices for reduction of environmental impacts to achieve sustainable development goals. Journal of Environmental Management, 248, 109238. https://doi.org/10.1016/j.jenvman.2019.07.009

93. Rajaeifar, M. A., Ghanavati, H., Dashti, B. B., Heijungs, R., Aghbashlo, M., & Tabatabaei, M. (2017). Electricity generation and GHG emission reduction potentials through different municipal solid waste management technologies: A comparative review. Renewable and Sustainable Energy Reviews, 79, 414–439. https://doi.org/10.1016/j.rser.2017.04.109

94. Read, A. D., Hudgins, M., & Phillips, P. (2001). Perpetual landfilling through aeration of the waste mass; lessons from test cells in Georgia (USA). Waste Management, 21(7), 617–629. https://doi.org/10.1016/S0956-053X(00)00124-0

95. Reinhart, D. R. (1993). A Review Of Recent Studies On The Sources Of Hazardous Compounds Emitted From Solid Waste Landfills: A U.S. Experience. Waste Management & Research, 11(3), 257–268. https://doi.org/10.1006/wmre.1993.1025

96. Rogers, T. W., de los Reyes, F. L., III, Beckwith, W. J., & Borden, R. C. (2013). Power earth auger modification for waste extraction from

pit latrines. Journal of Water, Sanitation and Hygiene for Development, 4(1), 72–80. https://doi.org/10.2166/washdev.2013.183

97. Saravanan, A., Senthil Kumar, P., Jeevanantham, S., Karishma, S., Tajsabreen, B., Yaashikaa, P. R., & Reshma, B. (2021). Effective water/wastewater treatment methodologies for toxic pollutants removal: Processes and applications towards sustainable development. Chemosphere, 280, 130595. https://doi.org/10.1016/j.chemosphere.2021.130595
98. Scarlat, N., Fahl, F., & Dallemand, J.-F. (2019). Status and Opportunities for Energy Recovery from Municipal Solid Waste in Europe. Waste and Biomass Valorization, 10(9), 2425–2444. https://doi.org/10.1007/s12649-018-0297-7
99. Schloemann, E. (1982). Eddy-current techniques for segregating nonferrous metals from waste. Conservation & Recycling, 5(2), 149–162. https://doi.org/10.1016/0361-3658(82)90024-8
100. Shafizadeh-Moghadam, H., & Helbich, M. (2015). Spatiotemporal variability of urban growth factors: A global and local perspective on the megacity of Mumbai. International Journal of Applied Earth Observation and Geoinformation, 35, 187–198. https://doi.org/10.1016/j.jag.2014.08.013
101. Shahidul Islam, Md., & Tanaka, M. (2004). Impacts of pollution on coastal and marine ecosystems including coastal and marine fisheries and approach for management: A review and synthesis. Marine Pollution Bulletin, 48(7), 624–649. https://doi.org/10.1016/j.marpolbul.2003.12.004
102. Sharholy, M., Ahmad, K., Mahmood, G., & Trivedi, R. C. (2008). Municipal solid waste management in Indian cities – A review. Waste Management, 28(2), 459–467. https://doi.org/10.1016/j.wasman.2007.02.008
103. Sharma, V. K., Fortuna, F., Mincarini, M., Berillo, M., & Cornacchia, G. (2000). Disposal of waste tyres for energy recovery and safe environment. Applied Energy, 65(1), 381–394. https://doi.org/10.1016/S0306-2619(99)00085-9
104. Shekdar, A. V. (2009). Sustainable solid waste management: An integrated approach for Asian countries. Waste Management, 29(4), 1438–1448. https://doi.org/10.1016/j.wasman.2008.08.025
105. Shiralipour, A., McConnell, D. B., & Smith, W. H. (1992). Uses and

benefits of MSW compost: A review and an assessment. Biomass and Bioenergy, 3(3), 267–279. https://doi.org/10.1016/0961-9534(92)90031-K

106. Siddiqua, A., Hahladakis, J. N., & Al-Attiya, W. A. K. A. (2022). An overview of the environmental pollution and health effects associated with waste landfilling and open dumping. Environmental Science and Pollution Research, 29(39), 58514–58536. https://doi.org/10.1007/s11356-022-21578-z
107. Singh, R. P., Tyagi, V. V., Allen, T., Ibrahim, M. H., & Kothari, R. (2011). An overview for exploring the possibilities of energy generation from municipal solid waste (MSW) in Indian scenario. Renewable and Sustainable Energy Reviews, 15(9), 4797–4808. https://doi.org/10.1016/j.rser.2011.07.071
108. Spokas, K., Bogner, J., Chanton, J. P., Morcet, M., Aran, C., Graff, C., … Hebe, I. (2006). Methane mass balance at three landfill sites: What is the efficiency of capture by gas collection systems? Waste Management, 26(5), 516–525. https://doi.org/10.1016/j.wasman.2005.07.021
109. Subramanian, P. M. (2000). Plastics recycling and waste management in the US. Resources, Conservation and Recycling, 28(3), 253–263. https://doi.org/10.1016/S0921-3449(99)00049-X
110. Sun, Y., Li, H., Lei, S., Semple, K. T., Coulon, F., Hu, Q., … Jones, K. C. (2022). Redevelopment of urban brownfield sites in China: Motivation, history, policies and improved management. Eco-Environment & Health, 1(2), 63–72. https://doi.org/10.1016/j.eehl.2022.04.005
111. Tai, J., Zhang, W., Che, Y., & Feng, D. (2011). Municipal solid waste source-separated collection in China: A comparative analysis. Waste Management, 31(8), 1673–1682. https://doi.org/10.1016/j.wasman.2011.03.014
112. Tilman, D., Cassman, K. G., Matson, P. A., Naylor, R., & Polasky, S. (2002). Agricultural sustainability and intensive production practices. Nature, 418(6898), 671–677. https://doi.org/10.1038/nature01014
113. Tin, A. M., Wise, D. L., Su, W.-H., Reutergardh, L., & Lee, S.-K. (1995). Cost—Benefit analysis of the municipal solid waste collection system in Yangon, Myanmar. Resources, Conservation and Recycling, 14(2), 103–131. https://doi.org/10.1016/S0921-3449(95)80004-2
114. Trois, C., Abera, Y., Dell'Orto, A., Kissoon, S., & Gómez-Sanabria, A. (2023). Application of the WROSE Model for Promoting Effective

Decision-Making and Sustained Climate Change Stabilization in the South African Waste Sector. In H. El Bari & C. Trois (Eds.), Waste Management in Developing Countries (pp. 39–62). Cham: Springer International Publishing. https://doi.org/10.1007/978-3-031-28001-6_3

115. van Puijenbroek, P. J. T. M., Beusen, A. H. W., & Bouwman, A. F. (2019). Global nitrogen and phosphorus in urban waste water based on the Shared Socio-economic pathways. Journal of Environmental Management, 231, 446–456. https://doi.org/10.1016/j.jenvman.2018.10.048
116. Vergara, S. E., & Tchobanoglous, G. (2012). Municipal Solid Waste and the Environment: A Global Perspective. Annual Review of Environment and Resources, 37(1), 277–309. https://doi.org/10.1146/annurev-environ-050511-122532
117. Vidanaarachchi, C. K., Yuen, S. T. S., & Pilapitiya, S. (2006). Municipal solid waste management in the Southern Province of Sri Lanka: Problems, issues and challenges. Waste Management, 26(8), 920–930. https://doi.org/10.1016/j.wasman.2005.09.013
118. Wang, X., & Gaustad, G. (2012). Prioritizing material recovery for end-of-life printed circuit boards. Waste Management, 32(10), 1903–1913. https://doi.org/10.1016/j.wasman.2012.05.005
119. Wang, Y., Gao, Y.-X., Song, J., Bonin, M., Guo, M., & Murphy, R. (2010). Assessment of technical and environmental performances of wheat-based foams in thermal packaging applications. Packaging Technology and Science, 23(7), 363–382. https://doi.org/10.1002/pts.897
120. Wen, Z., Hu, S., De Clercq, D., Beck, M. B., Zhang, H., Zhang, H., … Liu, J. (2018). Design, implementation, and evaluation of an Internet of Things (IoT) network system for restaurant food waste management. Waste Management, 73, 26–38. https://doi.org/10.1016/j.wasman.2017.11.054
121. Wilkinson, T. J. (1982). The Definition of Ancient Manured Zones by Means of Extensive Sherd-Sampling Techniques. Journal of Field Archaeology, 9(3), 323–333. https://doi.org/10.1179/009346982791504616
122. Wilson, D. C., Velis, C., & Cheeseman, C. (2006). Role of informal sector recycling in waste management in developing countries.

Habitat International, 30(4), 797–808. https://doi.org/10.1016/j.habitatint.2005.09.005

123. Wisner, B., & Uitto, J. (2009). Life on the Edge: Urban Social Vulnerability and Decentralized, Citizen-Based Disaster Risk Reduction in Four Large Cities of the Pacific Rim. In H. G. Brauch, Ú. O. Spring, J. Grin, C. Mesjasz, P. Kameri-Mbote, N. C. Behera, … H. Krummenacher (Eds.), Facing Global Environmental Change: Environmental, Human, Energy, Food, Health and Water Security Concepts (pp. 215–231). Berlin, Heidelberg: Springer. https://doi.org/10.1007/978-3-540-68488-6_13

124. Yellishetty, M., Ranjith, P. G., & Tharumarajah, A. (2010). Iron ore and steel production trends and material flows in the world: Is this really sustainable? Resources, Conservation and Recycling, 54(12), 1084–1094. https://doi.org/10.1016/j.resconrec.2010.03.003

125. Zhang, C., Guo, Y., Wang, X., & Chen, S. (2019). Temporal and spatial variation of greenhouse gas emissions from a limited-controlled landfill site. Environment International, 127, 387–394. https://doi.org/10.1016/j.envint.2019.03.052

126. Zhang, D. Q., Tan, S. K., & Gersberg, R. M. (2010). Municipal solid waste management in China: Status, problems and challenges. Journal of Environmental Management, 91(8), 1623–1633. https://doi.org/10.1016/j.jenvman.2010.03.012.

INDEX

M

N

O

P

Q

R

S

T

U

V

W